Science in Society

The world around us has been shaped by science and man's relationship to it and, in recent years, sociologists have been increasingly preoccupied with the latter. In *Science in Society*, Massimiano Bucchi provides a brief and approachable introduction to this sociological issue.

Without assuming any scientific background, Bucchi provides clear summaries of all the major theoretical positions within the sociology of science, using many fascinating examples to illustrate them. Theories covered include Thomas Kuhn's theory of scientific change, the sociology of scientific knowledge, actor-network theory, and the social construction of technology. The second half of the book goes on to look at some recent public controversies over the role of science in the modern world, including:

- the Sokal affair, otherwise known as the 'science wars';
- debates over public understanding of science, such as global warming and genetically modified food;
- the implications of the human genome project.

This highly readable text will be essential reading for all students studying the sociology of science.

Massimiano Bucchi teaches Sociology of Science in the Faculty of Sociology at the University of Trento, Italy.

Science in Society
An introduction to social studies of science

Massimiano Bucchi

Revised and expanded edition of *Scienza e Società. Introduzione alla sociologia della scienza*, Bologna: Il Mulino, 2002.
Translation by Adrian Belton.

LONDON AND NEW YORK

First published in Italian 2002
by Società editrice il Mulino, Bologna

First English language edition 2004
by Routledge
2 Park Square, Milton Park, Abingdon, Oxon OX14 4RN

Simultaneously published in the USA and Canada
by Routledge
270 Madison, Ave, New York NY 10016

Transferred to Digital Printing 2008

Routledge is an imprint of the Taylor & Francis Group, an informa business

© 2002 Società editrice il Mulino, Bologna
© 2004 Massimiano Bucchi, and Adrian Belton for the translation

Typeset in Times by
Florence Production Ltd, Stoodleigh, Devon
Printed and bound in Great Britain by
TJI Digital, Padstow, Cornwall

All rights reserved. No part of this book may be reprinted or reproduced or utilized in any form or by any electronic, mechanical, or other means, now known or hereafter invented, including photocopying and recording, or in any information storage or retrieval system, without permission in writing from the publishers.

British Library Cataloguing in Publication Data
A catalogue record for this book is available
from the British Library

Library of Congress Cataloging in Publication Data
Bucchi, Massimiano, 1970–
  [Scienza e società. English]
  Science in society: an introduction to social studies of science/
Massimiano Bucchi; translated by Adrian Belton.
    p. cm.
  Includes bibliographical references and index.
  1. Science–Social aspects. I. Title.
  Q175.5.B8316 2004
  303.48′3–dc22   2003017193

ISBN 10: 0-415-32199-9 (hbk)
ISBN 10: 0-415-32200-6 (pbk)
ISBN 13: 978-0-415-32199-0 (hbk)
ISBN 13: 978-0-415-32200-3 (pbk)

Contents

	Introduction	1
	Prologue	5
1	The development of modern science and the birth of the sociology of science	7
2	Paradigms and styles of thought: a 'social window' on science?	25
3	Is mathematics socially shaped? The 'strong programme'	41
4	Inside the laboratory	61
5	Tearing bicycles and missiles apart: the sociology of technology	77
6	'Science wars'	93
7	Communicating science	107
8	A new science?	125
	Suggested further reading and interesting websites	143
	References	147
	Index of names	159

Bohr: Why were so many of [theoretical physicists] Jews? Because theoretical physics ... was always regarded in Germany as inferior to experimental physics, and the theoretical chairs and lectureships were the only ones the Jews could get.
Margrethe: Physics, yes? Physics.
Bohr: This is physics.
Margrethe: It's also politics.
Heisenberg: The two are sometimes painfully difficult to keep apart.
(Michael Frayn, *Copenhagen*)

The world can come to an end, the phoenix can rise from its ashes, the Colosseum can catch fire ... but Standard Transformer Oil B, 11-Extra, is what it is and remains what it is.
(Carlo Emilio Gadda, *That Awful Mess on Via Merulana*)

Introduction

Science is increasingly at the centre of public debate. The role of the scientific enterprise, its responsibilities, its relationships with the social, political, religious and economic institutions, and the legal and administrative measures required to regulate scientific discoveries and technological innovations: all these are issues that appear with ever greater prominence and urgency on the political and public agenda.

Not only does the sociology of science play a marginal role in the debate on these matters, but when it is called upon to testify, it is treated in overly simplistic terms. 'Relativism', 'constructivism' and 'anti-scientism' are the pejorative epithets most frequently used to dismiss an entire sector of inquiry – or conversely to elevate it to the status of an ideological challenge against research in its entirety or, in the most extreme cases, against the capacity of human beings to understand the reality that surrounds them.

It is not the intention – even less the presumption – of this book to conduct an apologia for the sociology of science or to rehabilitate it in the eyes of scientists and commentators. Its much more modest aim is to describe some of the main contributions that have distinguished the discipline over the past fifty-odd years. Its purpose is to show, as far as possible, that sociology of science has developed into a broad and diversified area of research, with a wealth of empirical studies, and an abundance of internal debates often conducted in no less lively polemic with the outside.

Dismissing the discipline out of hand – in the manner of those commentators who spring to the attack whenever they hear 'sociology' coupled with 'science' – does not, I submit, have much more sense than attributing a single, monolithic position to the whole of the philosophy of science and 'accepting' or 'rejecting' it on that basis alone, thereby ignoring the substantial differences among Nagel, Popper and

Feyerabend. The reader may be surprised to learn that, for instance, the sociology of science does not coincide with the notorious 'strong programme'; that sociologists themselves are deeply divided on themes like relativism or constructivism; or that the statement 'nature does not exist, everything is constructed by society' would only be endorsed by a tiny number of scholars working in the discipline. Or again that many of the theses of the contemporary sociology of science were first put forward – and often in even more radical form – by a medical doctor in the first half of the last century (Fleck, 1935).

Of course, the book can only provide a partial survey of its subject matter, one restricted to the themes or approaches most distinctive of the discipline. It omits, for instance, systematic analysis of research policies. This area of inquiry has now acquired the size and status of a sector in itself, yet the sociology of science's contribution to it has often been only marginal compared to that by other disciplines.

Compared to a more rigidly chronological treatment, or one proceeding author by author, the advantage of the theme-based organization used by the book is that it shows how sociologists of science have developed their discipline in close dialogue with scholars working in other fields: primarily historians and philosophers of science, but also anthropologists, economists, political scientists, engineers and natural scientists. Indeed, there are and have been numerous university departments and journals operating under the generic denomination of 'Science Studies' or 'Science and Technology Studies' (frequently abbreviated to STS), most notably the celebrated Science Studies Unit of Edinburgh.

Also deliberately excluded from the book are certain 'classical' authors on the sociology of knowledge, such as Durkheim, Marx and Mannheim, even though they are often cited by sociologists of science and used as authoritative points of reference.[1]

A final caveat. Works of a theoretical-general nature are the exception in the general panorama of STS – especially since the 1970s. Rather, empirical case studies, often minutely documented, are the rule. The book makes brief mention of some of these case studies, but obviously without claiming to do justice to their complexity, since this would often require extensive preliminary description of the scientific matters treated. My advice is that the reader should use these citations to decide the cases of greatest interest and as a stimulus to read the work mentioned in its entirety. In all cases I have cited the text in which the subject is treated most briefly and accessibly. A list of suggested readings and interesting websites is included for each chapter.

I wish to thank Mario Diani, Silvia Fargion, Massimo Mazzotti, Federico Neresini, Giuseppe Pellegrini and James Wiley for their encouragement and their long conversations with me about this book; Adrian Belton for his patient work of translation and revision; David Bloor, Susan Howard, Donald MacKenzie, Bruce Lewenstein and Renato Mazzolini for their careful reading of previous drafts of the manuscript and their valuable suggestions. I am indebted to the late David Edge and the secretary of the Science Studies Unit, Carol Tansley, for their helpfulness during my sojourn at the University of Edinburgh. Some of the topics dealt with by the book were presented and discussed during seminars held at the University of Trento and the *Collegium Helveticum* of the Polytechnic of Zurich; I thank all those present for their patience and comments, Helga Nowotny in particular. Finally, I am grateful to the late Robert K. Merton and Harriet Zuckerman for discussing some of the book's themes with me.

Massimiano Bucchi
October 2003

Note

1 For a brief selection of studies in this regard see Nowotny and Taschwer (1996).

Prologue

Every morning, a few minutes before nine o'clock, Markus goes to work: he puts on a protective suit and heads for the laboratory. But the laboratory where Markus does his research has a circumference of 27 kilometres. He uses a duty car to move from one part of the laboratory to another, crossing the border between Switzerland and France many times a day to do so. A special lift takes him to 100 metres under the ground, where he greets his colleagues: 16 of them, among their number, physicists, engineers and other technicians, with whom Markus communicates in two foreign languages. The apparatus necessary for the team's research is produced in 13 different countries. The experiment on which they are currently engaged will last several months, during which time several people will join and leave the group.

The scenario of a science fiction tale? Or a glimpse of the future? Neither, more simply this is the typical day for one of the more than 300 physicists working at CERN (Center for European Research in Nuclear Physics), the largest laboratory for particle physics and the biggest experimental machine in the world. Staffed by physicists, engineers, technicians, manual workers and administrative personnel, the laboratory has a total of 3,000 employees and a budget that in 2000 exceeded 870 million Swiss francs (more than 500 million dollars) contributed by 20 member states (Austria, Belgium, Bulgaria, the Czech Republic, Denmark, Finland, France, Germany, Greece, Hungary, Italy, Luxembourg, the Netherlands, Norway, Portugal, Slovakia, Spain, Sweden, Switzerland and the UK).

What promises of economic, technological and military benefit does such a huge organizational and financial undertaking hold out? 'None. This is the most interesting thing.' The head of public relations at CERN smiles as he replies to the question by one of the many groups of visitors. 'What we do here is almost a mystical enterprise, almost a religion. It has no practical pay-off. It aims to gain understanding of where we come from, what matter is really made of'.

It is for this purpose that the huge accelerator used by the CERN physicists was built: to recreate conditions similar to those that existed at the beginnings of the universe, in order to understand how and where everything originated. Twenty countries, 500 billion dollars a year and almost 3,000 people working every day on one of the most sophisticated and abstract enterprises ever pursued by mankind.

Every year 60,000 visitors, mostly students, visit CERN, where they are welcomed by an efficient public visits service. They do not come to watch an experiment, as one might expect, nor to touch test tubes or feel inclined planes with their hands. For CERN experiments belong to the realm of the infinitely small and invisible. Moreover, in the period when experiments are under way (from spring to late summer) it is not possible to visit the accelerator. What, then, do visitors see? Huge tubes of incredible length, tangles of wires and computers as big as a bedroom. But obviously, the visitors have faith. They know that at some point, beyond their power of sight, the machines will make something happen and will record it and measure it.

It may be that CERN's head of public relations was right: it may be that a visit to CERN is no different from a pilgrimage to a sanctuary by the faithful, who do not expect to see and touch their God but know that He has revealed Himself in that place in the past and will do so again.

There is probably no better way to explain what science means today, to account for its importance in society and culture. We do not expect science only to turn on the lights in our homes or keep our food fresh. We want it to answer our most profound questions. This is perhaps the only feature shared by the science of Tycho Brahe – who made all his observations with the naked eye – and the science of Markus. Everything else has changed, beginning with forms and sizes.

1 The development of modern science and the birth of the sociology of science

1 From 'little science' to 'big science'

In 1963 a historian of science, Derek de Solla Price, published a short book in which he outlined the historic evolution of science and, in doing so, laid the foundations for the subject today known as scientometrics: the quantitative analysis of scientific activity that uses such indicators as the number of research papers, publications and citations (Price, 1963). Using very simple data, Price showed that the growth rate of scientific research during the past two centuries has been higher than that of any other human activity. One of the facts cited by Price, which later became proverbial, was that approximately 87 per cent of all the scientists who had ever lived were at work in the 1960s. The total number of researchers had risen from 50,000 at the end of the nineteenth century to more than one million. Similarly, the number of scientific journals had burgeoned from around 100 in 1830 to several tens of thousands; the proportion of Gross National Product devoted to scientific research in the US had risen from 0.2 per cent in 1929 to 3 per cent in the early 1960s. Science had also become a collaborative, as opposed to individual, enterprise: between the 1920s and 1950s, the percentage of scientific papers written by a single researcher published in American specialist journals diminished by half, while the ratio of papers signed by at least four researchers increased concomitantly (Klaw, 1968; Zuckerman, 1977).

In short, by the 1960s, artisan or 'little science' had become a huge enterprise in both social and economic terms. Physicist Alvin Weinberg termed this 'big science' in analogy with 'big business' – the great conglomerates of capitalist industry which grow exponentially and double in size approximately every 15 years. To give an idea of the pace of this growth, Price compared it with other phenomena, for instance the earth's population, which took around 50 years to double:

8 *Birth of the sociology of science*

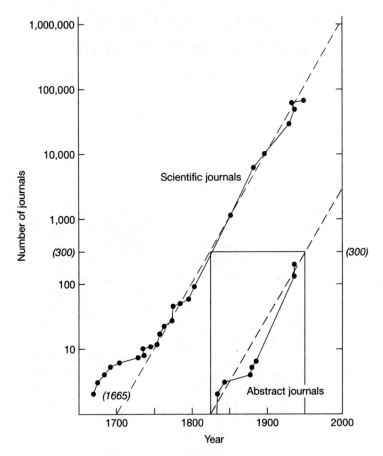

Figure 1.1 Total number scientific journals and abstract journals founded, as a function of date

Source: Price (1963)

The immediacy of science needs a comparison of this sort before one can realize that it implies an explosion of science dwarfing that of the population, and indeed all other explosions of non-scientific human growth. Roughly speaking, every doubling of the population has produced at least three doublings of the number of scientists.

Price based two interesting considerations on these data. First, he pointed out that the often emphasized role of the Second World War

Birth of the sociology of science 9

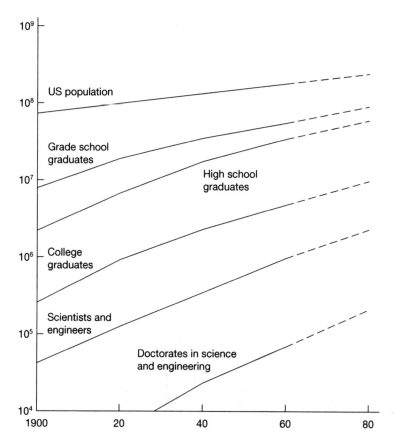

Figure 1.2 Growth of scientific manpower and general population in the US
Source: Price (1963)

in the development of scientific activity had largely to be reappraised. The growth rate had in fact remained stable in the years immediately after the war compared to the years immediately before it. If indeed the conflict had exerted any effect, it was a slight flattening in the growth curve due to the communication restrictions imposed on scientists by the exigencies of military secrecy. Price's second consideration was in fact a forecast. Unless a dramatic rearrangement took place, the exponential growth of science would inevitably encounter an upper limit. This saturation level, thought Price, would be reached more quickly in those countries – the US, for example, or the European states – where the increase in scientific activity had been

10 Birth of the sociology of science

in progress for longer, leaving margins for growth in countries, like Japan, of more recent scientific development. Price concluded:

> It is clear that we cannot go up another two orders of magnitude as we have climbed the last five. If we did, we should have two scientists for every man, woman, child, and dog in the population, and we should spend on them twice as much money as we had. Scientific doomsday is therefore less than a century distant.
>
> (Price, 1963: 17)

Experts and policy makers suggested various measures to deal with this exponential growth. For instance, Lord Bowden, at that time British Minister of Education and Scientific Research, proposed that restrictions should be set on the amount of money spent on the various research disciplines.

Since the 1960s, however, the development of science seems to have reached saturation point: the curve has levelled out, especially in terms of spending, and it has settled in most Western countries at

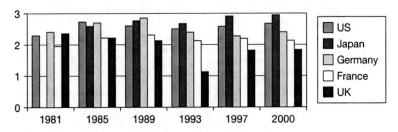

Figure 1.3 R&D expenditure as a percentage of GNP in some countries, 1981–2000

Source: Elaboration on NSB data, 2000; OECD data, 2002

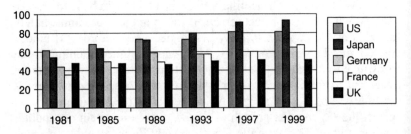

Figure 1.4 Number of researchers per 10,000 manpower units, 1981–1999

Source: Elaboration on OECD data, 2002

between 2 and 3 per cent of Gross National Product. But growth has, instead, continued in other areas of the world, Asia in particular. By the early 1990s, Japan had already overtaken the US in terms of its number of active scientists and engineers. Scientific research and technological development today involve approximately 3.4 million researchers in the OECD countries, for a total expenditure of around US$ 602 billion (OECD, 2002).

Price highlighted other features of contemporary science as well, for instance the 'immediacy effect' or the rapid obsolescence of specialized publications. Papers – i.e. the scientific articles that have become the communication medium of contemporary science, taking the place of the treatises or letters that scholars once used to address the scientific institutions because they allow the faster processing of discovery claims – tend to be cited very frequently in the period immediately following their publication. Thereafter, the citations rapidly diminish, disappearing completely after a period that on average is five years (although in sectors like physics and biomedical sciences the period is even shorter, around three years).

While it is relatively easy to trace recent developments in the curve representing scientific research, it is more difficult to identify the origin of that curve, or in other words, the beginning of the set of activities and institutions that we today call 'science'.[1]

Science historians agree that this period began between the mid-sixteenth and late seventeenth centuries, during the so-called 'scientific revolution'. Perhaps the most significant innovations brought by the latter to styles of thought and inquiry into nature were the following:

a   the adoption of distinctive methods and procedures for scientific activity, primarily experimentation;
b   the non-hierarchical character of knowledge. The scholar was no longer bound to accept 'by fiat' what his predecessors had produced; instead, he was encouraged to analyse it directly on his own. *De Humani Corporis Fabrica* by Vesalius (1543), for instance, includes a table with descriptions of all the tools required to dissect a body;
c   the demise of a teleological, man-centred cosmology and extensive discussion of the most appropriate methods with which to study nature;
d   the importance given to communication and the exchange of results and hypotheses – as opposed to the secrecy with which magical and alchemical works were shrouded – and the formation

of a 'scientific community' with specific arenas for discussion (the scientific academies founded since the seventeenth century, the journals devoted to the publication of results).[2]

This is not to imply that all the ideas and the practical and conceptual tools employed were radically new: anticipations of atomic theory or of heliocentrism, for instance, can be traced back to Ancient Greece (Butterfield, 1958). However, it was with the scientific revolution that these concepts to a large extent became the shared heritage of educated social groups. This growth and transformation of scientific activity was manifest in such events as the founding of the first academies and national science societies like the *Accademia dei Lincei* (1603), the *Accademia del Cimento* (1651), the *Royal Society* (1662) and the *Académie des Sciences* (1666). Scholars thus began to recognize each other and present themselves to the rest of society as a homogeneous community. They adopted internal rules and received external recognition of the importance and dignity of their role in society.

The processes of professionalization and institutionalization continued in the centuries that followed, with increasingly precise definition being given to the professional figure and social role of the 'scientist', a term first used by William Whevell in 1833 to describe the participants at a meeting of the British Association for the Advancement of Science. During the course of the nineteenth century, scientific practice found its natural setting in laboratories established on a permanent basis – for instance, the Cavendish Laboratory founded in Cambridge in 1871 and directed by physicist James Clerk Maxwell, the Museum of Comparative Zoology at Harvard and the Institut Pasteur in Paris. These laboratories further emphasized the differentiation among the scientific disciplines (and also among the sub-disciplines which are today the most common areas of endeavour for researchers), and among their relative communities, journals and forums, all of which were markedly international compared to other social activities. Since the scientific revolution, scientists have used a *lingua franca* – initially Latin, later French and English – to communicate with each other.

During the nineteenth century, the majority of the Western countries sought to emulate the organization of universities in Prussia, with their disciplinary specialization, their combination of teaching and research within the same institution, and their insistence on the 'academic scholar' left free to define the objectives and methods of his or her research (Ben-David and Zloczower, 1962; Ben-David, 1971).

Historians and sociologists have linked the institutionalization of science with other processes, perhaps most notably with industrialization or capitalism. This does not imply that the contribution of science has amounted to no more than its ability to supply the tools or technical innovations necessary for economic development, for instance in the textiles industry. At another level, some scholars have pointed out the affinity between the freedom to interpret nature by means of experimentation and individual observation untrammelled by tradition and capitalistic individualism. Nor should one underestimate the importance of the dissolution of barriers between scholars and craftsmen in enabling abstract thought to be combined with empirical observation and technical skills. For Barry Barnes, 'Rapidly expanding commercial and industrial middle classes saw in the "scientific style", rather than theology or the bible, a vehicle of cultural and symbolic expression' (Barnes, 1985: 16).

In his doctoral thesis on science, technology and society in seventeenth-century England (1938), Robert K. Merton argued that the relationship between scientific practice and capitalism is only indirect. He related the institutional development of science instead to the diffusion of particular religious values, just as Max Weber had done in his analysis of the birth of capitalism (Weber, 1905). Using a variety of historical data – for instance concerning the activity of the Royal Society's members in its early decades – Merton showed not only that an increasing number of individuals from the British elite of the time devoted themselves to science, but also that a significant proportion of their work had no practical pay-off. Their desire to practise science must, therefore, have been driven by other motives. A systematic and methodical mentality, or rationalism; diligence in the empirical and individual study of nature as revealing the greatness of God; commitment to practical activities as a sign of one's own future salvation: these were all elements highly valued by Protestantism and, at the same time, powerful incentives for scientific inquiry. As Robert Boyle wrote in his will with reference to his fellow members of the Royal Society:

> Wishing them also a happy success in their laudable attempts, to discover the true Nature of the Works of God; and praying that they and all other Searchers into Physical Truths, may Cordially refer their Attainments to the glory of the Great Author of Nature, and to the Comfort of Mankind.
> (quoted in Merton, 1938, repr. 1973: 234)

Besides pointing out the link between Puritanism and science, Merton took pains to emphasize that the institutionalization of science and the social codification of the scientist's role did not only presuppose a series of methods and activities. Rather, they also required a nucleus of social elements: that is, values and norms able to found science as a social subsystem which related to the rest of society but was at the same time endowed with autonomy. Merton thought that a specific branch of sociology should concern itself with study of these features and, therefore, with the relationship between science and society. This was to be the sociology of science.

2 The birth of the sociology of science

It has been often noted that sociology discovered science as a specific object of inquiry somewhat belatedly (Merton, 1952). Although the first studies were produced in the late 1940s, it was only in 1978 that, for instance, the association of American sociologists created sections devoted to the sociology of science. In 1976, the journal *Science Studies* changed its name to *Social Studies of Science* and thus became the first specialized journal in this disciplinary area.

Why so late? Robert K. Merton – later unanimously regarded as the founder of this sector of sociology – thought that the delay was at least partly due to scant awareness of the social role of science even in a country like the US (Merton, 1952). A first watershed came with the Second World War. From the early post-war years onwards, a variety of factors strengthened the belief that political power had grown increasingly dependent on the contributions of science and technology, and that the economic, social and ecological consequences of scientific discovery and technological innovation exerted a decisive influence on the fate of nations. The role performed by scientists and research teams during the First and, particularly, the Second World Wars was indubitably of great importance. The development and application of radar technology by the British armed forces, and the Manhattan Project which produced America's first nuclear weapons, are only two of the most significant examples of the close integration that came about in those years among political institutions, the military apparatus and researchers, especially in sectors such as physics. In Great Britain, two physicists, Henry Tizard and F.A. Lindemann, were respectively chief adviser to the Minister of Aeronautics and personal adviser to Prime Minister Churchill during the war (Snow, 1960). The new political equilibrium and configuration of international relationships that came about after the war were further factors. The government of the Soviet Union, America's main rival as superpower, had in fact already committed enormous resources to scientific research in previous decades. So impressed were Western intellectuals like John Bernal by this commitment that they declared it the model to emulate, arguing that the political supervision of research was crucial for the development of society (Bernal, 1939).

By the end of the Second World War, therefore, most of the industrialized countries had recognized that the state's active intervention in research was both possible and important. This conviction led to the creation of scientific committees to advise governments on policy objectives, both at the general level and within specific sectors, and also to the allocation of larger amounts of resources in pursuit of those objectives. 'A new breed of experts came into being, bridging the values and norm systems of the state and the academy, creating a new vocabulary and a new kind of role as science and technology politicians' (Eltzinga and Jamison, 1995: 582).

A further event of great significance took place in 1957. In that year, the Soviet Union launched the first artificial satellite in history; a feat which had enormous impact in the Western countries, and especially in the US, where the launch was considered indicative of the progress – and therefore the dangerous potential – achieved in science and technology by the rival superpower. The 'Sputnik effect' triggered reactions at two levels. First, there was a further expansion in research expenditure in the US, which grew by around 15 per cent per year until the early 1960s. At the same time, politicians and scientists became convinced that competition with the Soviet Union could only be sustained if greater commitment was made to university education and, in particular, to the training and recruitment of advanced researchers and technicians.

There were other reasons for the 'general neglect of the sociology of science' (Merton, 1952). Science had been traditionally considered an enterprise distinct from other human activities (such as industrial work, the formation of social and political movements, the diffusion of religious beliefs). It was protected by some sort of sacred aura and was therefore not susceptible to sociological inquiry.

But this was despite the fact that many sociologists had already begun analysis of the relationships between social conditions and cognitive activity in the thematic area known as the sociology of knowledge. Karl Mannheim made a decisive contribution to such analysis by going beyond the Marxian principle that only 'erroneous beliefs' can be explained on the basis of interests and material relations. For Mannheim, ideology was not necessarily a 'false consciousness'; rather, it was a mode of thought related to the position of an individual or a group within society (by which he meant that they belonged not only to a social class in the Marxian sense but also to a certain social group, or even to a certain generation) (Mannheim, 1925). At the same time, however, Mannheim continued to accord special knowledge status to the natural sciences and mathematics. In these areas, he maintained, it was not possible to discern the influence of social elements instead identifiable in philosophical and religious thought, or in artistic expression.

Prior to Merton, therefore, no explicit mention was made of a 'sociology of science'. Yet Merton was only the best-known representative of a group of American scholars who produced a series of studies on science and, in particular, on the institutional mechanisms governing science, from the 1950s onwards. Given this interest, the approach of this group has been given the general label of 'institutional sociology of science' (Hess, 1997). Many of these scholars had come to the study of scientific activity from the sociology of professions – and therefore considered science to be primarily an occupation – with a marked interest in social stratification and often using quantitative methods. Merton and the school which he formed at Columbia University played a key role in this group. Merton had studied at Harvard University in the 1930s with the historian of science George Sarton – the founder of *Isis*, one of the leading journals in the field – and the sociologist Pitirim A. Sorokin. His already-mentioned doctoral thesis on 'Science Technology and Society in Seventeenth Century England' adopted a solidly Weberian approach in its demonstration that puritan ethics had favoured the emergence of values crucial to the development of the scientific enterprise: namely, rationalism, individualism and empiricism.

Bringing science within the range of sociological inquiry had its price, however, for the content of scientific practice could not be subjected to the same type of scrutiny. The studies by Merton and his colleagues were therefore mainly concerned with the organizational and functional aspects of science as an institution capable of self regulation. But for analytical purposes such as these, examination of the very technical-scientific content was not regarded as more necessary than possession of medical skills in order to study the sociology of medicine, or theological expertise to study the sociology of religion.

This approach found its most significant – or at least most famous – expression in the description of the 'normative structure of science'. Which values and norms, Merton wondered, actually guarantee the functioning of science? He centred his answer on four 'institutional imperatives': (1) Universalism, (2) 'Communism', (3) Disinterestedness, (4) Organized scepticism (Merton, 1942).

1. *Universalism*: scientific claims and results are judged independently of the attributes of the individual who has advanced them, e.g. social class, race and religion. Scientists are rewarded exclusively on the basis of the results obtained.
2. *'Communism'*: results and discoveries are not the property of the individual researcher concerned, but belong to the scientific community and society at large. This imperative is based on the assumption that knowledge is the product of a collective and cumulative effort by the scientific community. The scientist does not obtain recognition for his/her activity if s/he does not publicize it and thus make it accessible to others.
3. *Disinterestedness*: every researcher pursues the primary objective of knowledge progress, indirectly achieving individual recognition.
4. *Organized scepticism*: every researcher must be willing to evaluate any hypothesis or result critically, including his/her own, suspending final judgement until all necessary confirmations have been obtained.

When enunciating these principles, Merton frequently stressed that they were to be considered valid from the institutional point of view, not in terms of the individual motivations of the scientist. In other words, he was not so naïve as to assume that scientists, just because they are scientists, possess greater moral stature than other professionals. However, he believed that the functionality of these norms to the science subsystem was proved by the critical reaction by the scientific community to those who deviate from them, as well as by the sanctions that it imposes. The existence of concrete behaviours deviating from these imperatives does not question them as such, just as a theft does not question the value of private property. Besides, if all behaviours actually did conform to norms, the latter would not be necessary.

This aspect of Merton's study of science has been criticized on the ground that it reflects the paradigm of a traditionalist approach which the sociological analysis of science must overcome. Such criticisms have viewed Merton's description of the normative structure of science as an idealization more prescriptive than descriptive in its intent. Like certain philosophers of science, it is alleged, Merton presented science as it should be, rather than trying to depict scientific activity as it really is.[3] Using a series of case studies, some scholars have consequently tried to show the discrepancy between Merton's theory and the actual behaviour of scientists (Barnes and Dolby, 1970).

Merton, in fact, revised his original formulation by developing the concept of 'sociological ambivalence' to describe the situation of certain social actors – including scientists – when they must deal with conflicts among diverse values, norms and roles (Merton, 1963). In the early 1970s, several studies, including a detailed survey of forty-two scientists involved in analysis of data on the moon's surface collected by the Apollo missions, sought to demonstrate that such ambivalence gave rise to a 'dynamic alternation of norms and counter-norms' (Merton and Barber, 1963: 104). The institutional imperatives enunciated by Merton were thus matched by counter-norms such as 'particularism, interestedness and organized dogmatism' (Mitroff, 1974). The scientists interviewed by Mitroff, for instance, attributed

Table 1.1 Norms and counter-norms in science

Norms	Counter-norms
Universalism Scientific claims and findings are judged independently of the personal or social attributes of their proponents: social class, race, religion.	*Particularism* A scientist's social characteristics are factors which importantly influence how his/her work will be judged.
'Communism' Findings and discoveries are not the property of the individual researcher but belong to the scientific community and to society at large.	*Individualism* Property rights are extended to include protective control over results.
Disinterestedness Scientists pursue their primary aim, knowledge progress and indirectly achieve individual rewards.	*Interestedness* The individual researcher seeks to serve his/her own interests and those of the restricted group of scientists to which s/he belongs.
Organized scepticism Every researcher is obliged to scrutinize every hypothesis or finding carefully, including his own, suspending final judgement until the necessary confirmations become available.	*Organized dogmatism* The scientist must believe in his/her own findings with utter conviction while doubting those of others.

Source: Adapted from Mitroff (1974)

the following characteristics to themselves and to their colleagues: a reluctance to make certain aspects of their research public; an attachment to their own hypotheses; an unwillingness to abandon these hypotheses even in the presence of data contrary to them; or the tendency to judge results and claims on the basis of the social attributes (nationality, academic position) of the scientist advancing them. However, these counter-norms – as the interviewees themselves recognized – may play a positive role in scientific inquiry. Judging a researcher on his/her personal characteristics instead of his/her results, for instance, may save time because it focuses attention on work by scientists more likely to deliver results. Sticking to one's hypotheses may discourage the premature abandonment of research that might prove – at least indirectly – fruitful in the long run. Finally, the counter-norm of 'secrecy' may protect the scientific community against paralysis due to disputes concerning priority on a certain discovery[4] or pressures applied by the government or public opinion.

It was clearly unlikely that either Merton's imperatives or the specular counter-norms proposed by Mitroff could accurately describe the concrete behaviour of every scientist. A possible alternative was to regard both sets of imperatives as flexible 'ideological-rhetorical' repertoires from which scientists might draw from time to time and, according to the situation, in order to make sense of their actions and account for them to colleagues, policy makers and public opinion (Mulkay, 1979). For instance, in certain circumstances, secrecy can be condemned as misconduct towards other scientists; in other cases, it can be justified by the need for more accurate verification of one's findings before publishing them. Presentation of a discovery at a press conference before the official article has been published in a scientific journal may be greeted benevolently – as it was when 'wrinkles' in Cosmic Background Radiation were discovered by a team of NASA astronomers (Miller, 1994) – or harshly criticized as misconduct – as in the case of the alleged discovery of 'cold fusion' (Lewenstein, 1992a).

3 The Matthew effect and the forty-first chair: competition and inequality in science

How science works and the norms that regulate it have been analysed in great detail by Merton, and also by other scholars working within the institutional approach, notably Bernard Barber, Harriet Zuckerman and Warren Hagstrom. Hagstrom, in particular, wrote a classic study on the way in which the scientific community functions, highlighting the importance of recognition as a motivating factor (Hagstrom, 1965). One of the principles which Hagstrom deemed of particular relevance in this regard was, in fact, the anthropological principle of gift-giving, which significantly structures relationships in the scientific community. Thus, the researcher gives (and not sells) his paper to the journal in which it is published, gives copies of his or her papers to colleagues, and more generally offers his or her findings to the scientific community, which acknowledges their value by accepting them. This act of freely donating one's work is, for Hagstrom, symbolized by the rich mythology that describes scientists as unable to profit from their success during their lifetimes but, nevertheless, satisfied by leaving it for posterity. The case of Copernicus is emblematic: he received a printed copy of his book *De Revolutionibus* only when he was on his deathbed.

Merton and his closest collaborators – the so-called 'first circle' – produced a number of studies on how resources and rewards (such as opportunities to publish, or prestige) are assigned and distributed within the scientific community. Merton gave the name of the 'Matthew effect' to one of the phenomena that he observed (Merton, 1968a). The expression comes from the passage in Matthew's Gospel which runs: 'For unto every one that hath shall be given, and he shall have abundance: but from him that hath not shall be taken away even that which he hath'. In science this principle translates into a cumulative effect which exponentially rewards those who already occupy a privileged position. 'A scientific contribution will have greater visibility in the community of scientists when it is introduced by a scientist of high rank than when it is introduced by one who has not yet made his mark' (Merton, 1968a: 447), or – to use the words of a Nobel laureate in physics – '(the world) tends to give credit to already famous people' (cited in Merton, 1968a: 443).

On analysing empirical data, Merton and his group found, for instance, that papers submitted to a scientific journal were accepted more frequently if one of the authors was a Nobel prize-winner or a particularly well-known researcher. Similarly, a scientist's papers were cited much more frequently after he had received some highly visible reward like the Nobel prize. As a paradigmatic case, Merton cites an episode involving Lord Rayleigh, the great physicist. Rayleigh's name had been accidentally removed from a paper submitted to the British Association for the Advancement of Science. The association committee rejected the paper, judging it to be 'the work of one of those curious persons called paradoxers' (Merton, 1968a: 457). As soon as the name of the real author was disclosed, however, the paper was accepted. In 1996, a few hours after his application for government funding to continue studies on the structure of a new form of carbon named C_{60} had been turned down by a research council, the chemist Harry Kroto was announced as the winner of the Nobel prize for those same studies. The announcement led the research council to immediately reverse its decision (Gregory and Miller, 1998). Merton considers these mechanisms to be due to the scarcity of 'recognition' as a resource, and to rigidity in the forms of its allocation. The same thing may happen in science – especially as regards its greatest honours like the Nobel prize – as occurred at the Académie Française, which only had forty chairs. Among those relegated to the 'forty-first chair', i.e. the famous men excluded, were Descartes, Pascal, Rousseau, Diderot, Stendhal, Flaubert, Zola and Proust.

Merton considers the Matthew effect 'dysfunctional for the careers of individual scientists, who are penalized in the early stages of their development' but functional for the scientific system, insofar as it allows rapid selection to be made from the huge amount of papers submitted to journals. In certain cases, the names of highly visible scientists are able to direct the attention of the community to particularly innovative findings that would otherwise be ignored.

A quantitative measure of this tendency to elitism in science has been provided by Price, with the law that bears his name: 'half of the scientific papers published in a given sector are signed by the square root of the total number of scientific authors in that field'.

In other words, a relatively small number of highly productive researchers are responsible for most publications (Hess, 1997). Both Price's law and the Matthew effect depict the scientific community as a structure characterized by marked inequality and a heavily pyramidal distribution of resources (and especially of rewards: research funds, opportunities to publish, prizes and awards). Moreover, inequality and concentration of rewards tend to perpetuate and reinforce themselves over time.

The institutional approach has been used to analyse several other similar mechanisms. The 'halo effect', for instance, works to the advantage of scientists in more favourable institutional positions: a post at particularly prestigious university or department, for instance, or a particularly important role within the institution (Crane, 1967). According to a study by Barber, in 1962, 38 per cent of all US federal funds for research were assigned to only ten institutions and 59 per cent of all funds to only twenty-five (Klaw, 1968). More recent studies have argued that these mechanisms have an even more marked discrimination against participation by women in scientific activity. Vera Rubin, an astronomer who later in the 1970s was the first to question the thesis of regular expansion of the Universe, wrote in 1948 to Princeton University asking to apply for its Ph.D. programme in astrophysics, but received no reply. Women were admitted to Princeton only in 1971; and at the beginning of the 1980s, women still represented less than one quarter of the US's scientific population. Another case often mentioned is that of Rosalind Franklin, who never received recognition for her significant contribution to the discovery of DNA structure, for which James Watson and Francis Crick received the Nobel prize in 1962. Rossiter has called this discriminatory mechanism the 'Matilda effect' after the nineteenth-century writer and feminist activist, Matilda Gage, who authored an essay in which she claimed that the cotton gin was invented by a woman.

Although the bulk of the sociological literature on science has latterly developed in more or less explicit opposition to the institutional approach, several concepts and terms now widely employed were first introduced by Merton and other authors in his tradition. An example is the term 'gatekeeper', coined to indicate those scientists or subjects who, because they occupy a key position within a scientific institution, are able to influence the distribution of resources like research funds, teaching positions or publishing opportunities. Or the term 'invisible colleges', introduced on the basis of a seventeenth-century expression to denote the informal communities of researchers that cluster around a project or a research theme and which often turn out to be more influential, in terms of knowledge production, than formal communities (departments, research centres, scientific committees). One feature of the institutional approach that instead diluted into subsequent 'schools' of sociology of science was its connection with general sociological theory and with key sociological concepts. Several stimuli offered by this approach, as well as by Merton's specific work, have been largely obscured by the heated discussion on the 'institutional imperatives' that he attributed to science. The critical attention paid to this aspect is explained by Merton himself – who never officially entered the debate – in terms of expectations by scholars at the time.

> It was conceptual, after many descriptive works by historians of science. The key was to analyse science *also* as a social institution. As an institution, science must have norms – just as political, economic and religious institutions cannot exist without norms.[5]

Later opposition to the institutional approach has also bred the image of a 'naively positivistic' Merton committed to an idealized picture of scientific activity. But this is a Merton who probably never existed; as suggested by the following quotation – which expresses a position not very distant from those later taken up by several post-Mertonian sociologists of science.

> [There is a] rock-bound difference between the finished versions of scientific work as they appear in print and the actual course of inquiry followed by the inquirer. The difference is a little like that between textbooks of 'scientific method' and the ways in which scientists actually think, feel and go about their work. The books on method present ideal patterns: how scientists 'ought' to think, feel and act, but these tidy normative patterns ... do not reproduce the typically untidy, opportunistic adaptations that scientists make in the course of their inquiries. Typically, the scientific paper or monograph presents an immaculate appearance which reproduces little or nothing of the intuitive leaps, false starts, mistakes, loose ends, and happy accidents that actually cluttered up the inquiry.
> (Merton, 1968b: 4)

Notes

1 On the historical development of science see Ben-David (1971), Barnes (1985).
2 See for instance Cohen (1985), Hall (1983), Rossi (1988a, 1997).
3 To understand Merton's insistence on norms as functional imperatives and therefore on the capacity of science to regulate itself, one should bear in mind that he first approached the subject during the Second World War. This was a historical period when the essential features of science in a democratic society appeared, indeed, to be its autonomy and its ability to resist political, economic and religious pressures.
4 Merton (1973) finds highly revealing this type of controversy and, in particular, the cases when two or more researchers independently achieve the same discovery. Such cases, according to him, demonstrate the importance of taking into account the social and cultural dimension – beside individual creativity – when analysing scientific activity.
5 Personal communication, 5 April 2001.

2 Paradigms and styles of thought
A 'social window' on science?

1 A plant that divides botanists

The gilia is a genus of shrub belonging to the *Polemoniaceae* family which grows spontaneously in North America and is widely cultivated for its flowers. The genus comprises various species, one of which is the *Gilia inconspicua*. For some botanists the latter is a single species; for others it is a complex of fully five different ones. How is it possible for scientists to disagree so radically on something apparently as obvious and routine as the classification of a plant?

For explanation we must consider the *Gilia inconspicua* as one example among many of the antithesis between two approaches to plant classification. The first of them derives largely from Linnaeus' system and classifies species on the basis of the form and number of the sexual characteristics of plants, namely their stamens and pistils. It is a system that is easy to apply and of great practical utility – and it was especially so in the late 1700s and 1800s when classifying the huge number of newly discovered exotic plants arriving in Europe from the colonies was of vital importance. Moreover, the principle on which it was based – the principle that species are distinct entities separated by intrinsic differences – was in perfect accord with the biblical tradition. Thereafter, however, serious doubt was cast on that principle by Darwin's theory that species were 'abstractions, fictions of the taxonomist's mind rather than objectively existing entities in nature' (Dean, 1979: 216).

However, the Linnaean classification method was able to resist – with some small adjustments – the progressive spread and acceptance of the Darwinian theory. And it did so despite significant discoveries in botany-related disciplines during the early 1900s which greatly counselled against its use. In particular, the advances in genetics brought by rediscovery of Mendel's theories gave rise to a new classification method based on the experimental analysis of cells and their genetic heredity.

Yet, there was no wholesale abandonment by taxonomists of either the Linnaean system of nomenclature or of its procedures to describe, classify and name species of plants. The categories of the Linnaean system (class, order, genus and species) continued to be used in monographs, and taxonomists continued to insist that different species must display clear morphological differences. Some of these controversies still persist today and they have led to contrasting conclusions. *Gilia inconspicua* is a case in point, with experimental taxonomists distinguishing five species of the plant and morphologists insisting that there is only one.

2 Science and revolutions

The resistance raised by certain ideas, methods and instruments to change in their theoretical and experimental settings, their ability to survive (like the living beings studied by Darwin) by adapting to new findings or by selectively ignoring them, and their inevitable final extinction – and then the possibility of seeing (or not seeing) the same object in an entirely different manner by observing it with other conceptual and interpretative apparatuses – are themes obviously crucial to a sociology of scientific knowledge. They have been addressed in the celebrated work by the historian of science Thomas Kuhn, *The Structure of Scientific Revolutions* (Kuhn, 1962). This book, and the reading given to it by certain sociologists of science, gave significant impetus to study of the relationships between science and society.

In his book, Kuhn sets out a theory of scientific change grounded on a set of key concepts: those of 'paradigm', 'revolution' and 'normal science'. According to Kuhn, science does not advance smoothly along a linear path and by gradual approximations to the truth; rather, it is characterized by abrupt 'leaps' and profound 'discontinuities' – revolutions in a word. These discontinuities interrupt periods of 'quietness' characterized by the conduct of what Kuhn terms 'normal science'. On his definition, normal science is: 'Research firmly based upon one or more past scientific achievements, achievements that some particular scientific community acknowledges for a time as supplying the foundation for its further practice' (Kuhn, 1962: 10).

This outcome, or set of achievements, orients the work of scientists operating in a particular sector, and in a particular period, on the basis of generally untroubled consensus among them. Disputes may arise over who first made a certain discovery, wrangles may break out on specific aspects or the validity of particular experimental results, but there is no disagreement on fundamental issues.

Kuhn gives the name of 'paradigm' to this result or group of results. The Linnaean classification system is indubitably a paradigm. And so too, given that they are theories which have guided research for long periods in the history of science, are Ptolemaic astronomy, Copernican astronomy, Newtonian physics and Darwinian evolutionary theory.

An example of a recent and solidly founded paradigm is the Big Bang theory, or the idea that the universe originated in a specific event (a 'singularity'). Formulated in the 1940s, and considered proven by Penzias and Wilson's discovery of cosmic background radiation in 1965, this idea currently assumes the status of a paradigm. It shapes the research, experiments and observations of physicists and astronomers, providing them with a general framework into which they are able to fit even conflicting hypotheses – for example, on the evolution of the universe since the Big Bang, which they view as either uniform or an 'inflation' (i.e. an initial exponential expansion).

A paradigm is of key importance from a practical point of view as well, because it provides scientists with a solid basis from which to start without constantly having to 're-prove', or to argue from scratch, every aspect of their sector of inquiry.

> That can be left to the writer of textbooks. Given a textbook, however, the creative scientist can begin his research where it leaves off and thus concentrate exclusively upon the subtlest and most esoteric aspects of the natural phenomena that concern his group.
>
> (ibid.: 20)

From this point of view, according to Kuhn, the emergence of a paradigm signals that a research sector has consolidated itself into a scientific discipline. After the multiple views and schools of thought that characterize initial reflection on a particular theme, research and study stabilize around one of these schools or views.

During periods of normal science, therefore, a scientific community devotes its efforts mainly to refining and extending the paradigm in force, for example by obtaining more precise quantifications of certain variables. Hence, a paradigm furnishes the scientists with not only a reference theory but also an entire constellation of results, ideas and practices, examples, and standardized procedures with which to solve the problems that arise in the course of their research.

28 Paradigms and styles of thought

For Kuhn, normal science is an endeavour entirely similar to 'puzzle solving', a cumulative practice which aims to expand and consolidate the reference paradigm.

> The success of a paradigm ... is at the start largely a promise of success discoverable in selected and still incomplete examples. Normal science consists in the actualisation of that promise, an actualisation achieved by extending the knowledge of those facts that the paradigm displays as particularly revealing, by increasing the extent of the match between those facts and the paradigm's predictions, and by further articulation of the paradigm itself.
>
> (ibid.: 24)

This is by no means to imply that normal science is a second-class activity, humdrum and unimaginative, as one might believe when comparing it to the phases of a scientific revolution. Kuhn himself points out that normal science is quantitatively the largest part of scientific activity. Some of the most complex and costly scientific research projects of recent years, for example the Human Genome Project (to map human genes in their entirety, brought to conclusion in early 2001 by the HGP public consortium and the private company *Celera*), to a large extent belong to the domain of normal science, in that they seek to develop and complete knowledge founded on an already existing paradigm – in this case based on the structure of DNA discovered by James Watson and Francis Crick.

In 1989, when NASA began its COBE satellite mission intended to make accurate measurement of the ripples in cosmic background radiation – what astrophysicists believe to be the echoes of the original big bang – one of the scientists involved declared: 'I do not expect to overturn Big Bang theory with what we see, because it is a good theory and works well'.

If one examines the scientific analysis of a specific theme, therefore, one can observe a succession of paradigms. Light, for example, was studied until the end of the 1700s in terms of a Newtonian paradigm which regarded it as made of material corpuscles. Scientific research, therefore, sought to identify the movement of these particles and their impact on solid bodies. At the beginning of the 1800s, this paradigm was replaced by the conception of light as a wave movement. Finally, almost a century later, with the advent of quantum mechanics, light came to be regarded as composed of specific elements (photons) with mixed properties of wave and particle.

So how does the change from one paradigm to another come about? It is here that Kuhn introduces his concept of scientific revolution in antithesis to periods of normal science. Given the features of paradigms, which are much broader, more solid, and more intricately structured than a simple theory or assertion, Kuhn obviously does not believe that a result which is unexpected or contradictory to the paradigm is sufficient for the latter to be 'falsified' (as Popper instead maintained) and therefore discarded: 'to evoke crisis, (an anomaly) must usually be more than just an anomaly. There are always difficulties somewhere in the paradigm-nature fit' (Kuhn, 1962: 82).

In other words, the paradigm must be to some extent 'stretched' and shaped so that it can be defended against the attack of results and observations that contradict it. It is part of a paradigm's nature as a 'perceptive filter' to emphasize those features of reality that accord with it, and instead ignore those that gainsay it. To explain how this perceptive selection works, Kuhn cites an experiment conducted by a group of psychologists. The subjects were shown a series of playing cards in rapid sequence. Interspersed among the normal cards were a number of 'anomalous' ones: for example a red four of spades or a black seven of hearts. In the majority of cases the subjects classified these anomalous cards according to the usual categories. For example, the red four of spades was judged to be a four of hearts and the black seven of hearts to be a seven of clubs. When the exposure time was increased, some subjects began to realize that something was wrong. Yet many of them were unable to detect the anomalous cards even if they were given even forty times longer to do so. In some cases, the result was acute confusion and frustration, to the point that the subject was no longer able to say what the symbol for spades or hearts looked like.

A number of psychological and sociological studies of scientific work have highlighted similar mechanisms of 'confirmation bias', or the tendency to give greater importance to data which concur with one's theoretical model than to ones which do not. In an experiment conducted on two groups of scientists with different theories on the synthesis of ATP – a substance with a crucial role in the energy metabolism of living beings – the members of each group were asked to explain the results obtained by their own group and then those obtained by the rival one (Gilbert and Mulkay, 1982). It was found that when the scientists judged their own results, they considered them to be 'objective'; when they judged the results of others they said they were distorted by subjective bias. Another study, during which numerous researchers were interviewed, singled out different

30 Paradigms and styles of thought

types of reactions by scientists to data which conflict with their theoretical convictions (Chia, 1998). These responses are illustrated in Figure 2.1.

A scientist may or may not be able to 'see' an anomaly. If s/he does not see it, his/her response may be 'positivist' in that s/he clings to his/her theoretical convictions by simply ignoring the data in question (B). On the other hand, some of the scientists interviewed by Chia said that they did not need to come up against any contrary data to abandon a certain hypothesis (fideistic response, D). By way of example, consider the numerous physicians who embraced Einstein's theory of relativity well before it was confirmed by astronomical observations in 1919.

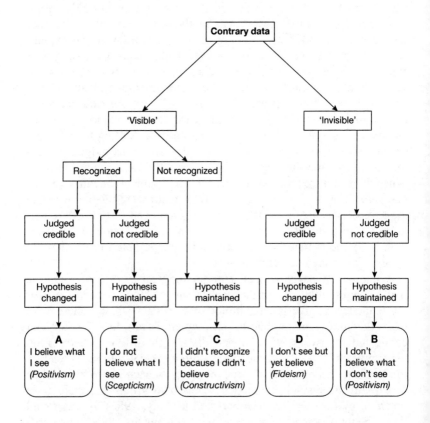

Figure 2.1 Types of response by scientists when faced with data contradicting their own theories

Source: Adapted from Chia (1998: 375)[1]

If the scientist 'sees' the anomaly, s/he may nevertheless not recognize it as such – some of the scientists interviewed acknowledged with hindsight that they had behaved in this manner because they were clinging to a particular interpretative scheme ('constructivist' attitude, C). In the case where the scientist 'sees' the anomaly and recognizes it, s/he may consider it to be credible (once again a positivist-type response which in this case induces the scientist to change his/her hypothesis, A) or not credible. In the latter case, s/he may maintain his/her hypothesis by adopting an attitude of scepticism. 'Observation does, in a sense, discipline theory in science, but theory disciplines observation also: observation reports may be discarded on theoretical grounds just as theories may be discarded on observational grounds' (Barnes, 1990: 63).

Kuhn concurs that it is not enough for a scientist to be confronted with data impossible to fit into the dominant paradigm framework for him/her to be induced to abandon it. The jettisoning of a paradigm, in fact, requires that there must be another one available to take its place: 'To reject one paradigm without simultaneously substituting another is to reject science itself' (Kuhn, 1962: 79).

A paradigm can be conceived as a castle whose layout and structure are well known, and whose every bastion is guarded by soldiers armed to the teeth. The breaching of a door, the crumbling of a wall or the collapse of a tower are not enough for the decision to be taken to abandon the castle. The threat must be particularly severe; and nearby there must be another castle of equal if not greater comfort to which the garrison can move.

> The transition from a paradigm in crisis to a new one from which a new tradition of normal science can emerge is far from a cumulative process, one achieved by an articulation or extension of the old paradigm. Rather it is a reconstruction of the field from new fundamentals, a reconstruction that changes some of the field's most elementary generalizations.
>
> (ibid.: 84)

To continue with the metaphor, it is not simply a matter of replacing one castle now fallen into disrepair with another of the same design but more modern, better equipped and more commodious. The transfer instead takes place between castles with radically different layouts and structures, and from whose towers territory with very different features can be policed.

> If there were but one set of scientific problems, one world within which to work on them, and one set of standards for their solution, paradigm competition might be settled more or less routinely by some process like counting the number of problems solved by each. But, in fact, these conditions are never met completely. The proponents of competing paradigms are always at least slightly at cross-purposes.
>
> (ibid.: 146–147)

For example, the paradigm of the 'continental drift' that followed the break-up of the original supercontinent, Pangaea, long laboured to gain acceptance among geologists, and this was because none of them had ever thought of collecting data on the motion of the continents when the old paradigm of *terra firma* held sway. When commenting on the resistance encountered by Wegener – who had formulated his theory of continental drift as early as 1915 although it was not generally endorsed until the 1950s – the geologist Du Toit pointed out that acceptance of the theory entailed 'the rewriting of numerous text-books, not only of Geology, but Palaeogeography, Palaeoclimatology and Geophysics' (Du Toit, 1937, cited in Cohen, 1985: 456). Scientists working from different theoretical perspectives have even been compared to natives from different tribes who find it impossible to communicate with each other (Feyerabend, 1975).

Hence, if it is not – or not only – a paradigm's match with reality that determines its supremacy, what is it that 'persuades' a group of researchers to abandon one paradigm and embrace another?

Kuhn does not provide an unequivocal explanation of the matter. However, he emphasizes that a shift between paradigms often corresponds to a change of generations. He cites a celebrated saying by the physicist Max Planck, 'A new scientific truth does not triumph by convincing its opponents and making them see the light, but rather because its opponents eventually die, and a new generation grows up that is familiar with it' (Kuhn, 1962: 150).

An old paradigm may be abandoned for numerous reasons, and not infrequently they are ones of an 'extra-scientific' nature. Consider, for instance, certain philosophical or religious beliefs. Kuhn cites the case of Kepler, whose conversion to Copernican theory was encouraged by his membership of a 'sun-worshipping cult'. He also mentions the importance of factors like the personal characteristics of the scientists propounding the new paradigm: their fame, their influence, even their nationality. Louis Pasteur, for example, during the scientific controversy that surrounded his attempt to explain the infective origin of certain pathologies, on several occasions invited his compatriots to reject rival theories on the ground that they were 'German' (Cadeddu, 1991). During the late 1950s and early 1960s, the cosmological debate on the origin of the universe polarized between the 'American' theory of the Big Bang and the 'English' one of the 'steady-state universe' – an opposition which earned the astronomer Fred Hoyle, one of the main proponents of the steady-state theory, the epithet of 'communist'.

The choice of the political metaphor of 'revolution'[2] to denote a change of paradigm is no coincidence, therefore. For the struggle between the supporters of different paradigms is a political struggle, and the winners raze the losers' castle to the ground: just as after the French Revolution names and calendars were changed, noble titles were abolished and 'history' was rewritten.

> It is rather as the professional community had been suddenly transported to another planet where familiar objects are seen in a different light and are joined by unfamiliar ones as well ... what were ducks in the scientist's world before the revolution are rabbits afterwards.
>
> (ibid.: 110)

3 Why is the cassowary not a bird?

Kuhn's book sparked wide debate among historians and philosophers of science, and it attracted numerous criticisms. According to one reviewer, indeed, Kuhn used the term 'paradigm' with fully twenty-two different meanings. But the large majority of sociologists saw his work as a chance to develop an analysis of the relationships between science and society which could overcome the limitations of the Mertonian approach.

What are paradigms, in fact, if not a means to convey social questions into scientific activity? The adoption of a paradigm and its maintenance involve mechanisms like authority, social control, trust and socialization (of young researchers and 'novices' in general). The paradigm furnishes the young researcher with a conventional problem-solving model, but knowing how to recognize the problems to which it is applicable, and the model's application itself, are based on consensus within a certain community.

Let us take another celebrated example from Kuhn. A little boy is walking with his father through a public garden. The father points to a bird, saying 'Look, there's a swan'. A little time later the boy points

34 *Paradigms and styles of thought*

to another bird: 'A swan', he says. His father corrects him, 'No, that's a duck'. Gradually, the boy learns which differences and similarities are significant and which are not: in other words, he is socialized to the type of classification which pertains to the community to which he belongs.

The hand that fits experimental data and results into a particular paradigm, or classifies them as anomalies, is a hand dressed in the characteristic clothes of a community or culture. Is the *Gilia inconspicua* only one species or five different species? Is mercury a metal even though it looks like a liquid? And why is a cassowary not a bird?

Why, then, is a cassowary not a bird? To answer we must briefly turn to the Karam tribe of New Guinea (Bulmer, 1967). The Karam use the term 'yakt' for numerous animals that we would classify as birds: parrots and canaries, for example. However, they also consider bats to be 'yakt' because they can fly, although we would classify them as mammals. But they do not consider 'kobtiy' (i.e. cassowaries) – furry

Figure 2.2 An image of the cassowary taken from the *Dizionario delle Scienze Naturali*, Florence, V. Battelli and Sons, 1830, vol. v, table 244, courtesy of Biblioteca Internazionale La Vigna

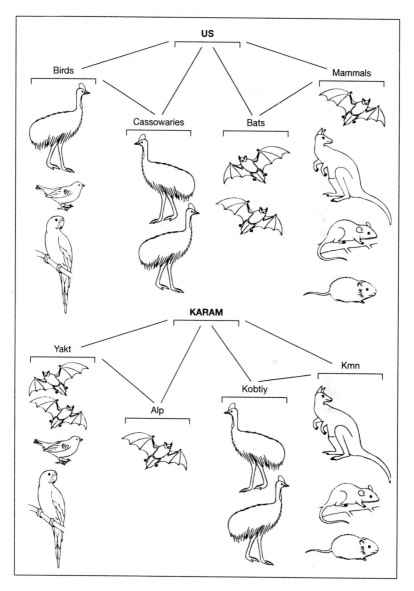

Figure 2.3 Scheme adapted from Barnes (1982a)

egg-laying bipeds of particular symbolic significance for the Karam – to be 'yakt'. Thus, asking 'why is the cassowary not a bird?' is like asking 'why for us are kobtiy yakt?' (Barnes, 1982a).

Both of these forms of classification are in accordance with experience. The classification used by the Karam seemed entirely workable to the anthropologist who studied it so carefully (Bulmer, 1967). The Karam lost no practical information, nor did they produce any inconsistency, by distinguishing between 'yakt' and 'kobtiy'. Both classifications – as devices which shape our routine knowledge and enable us to pigeon-hole 'anomalous' objects – should be backed by activities of social control and cultural transmission (Barnes, 1982b). For Barnes, no paradigm guarantees per se the rules for its automatic and appriate application. At first sight, the Karam classification might have induced anthropologists to believe that the category 'yakt' corresponded to our category 'birds', or at least they would have done so until they had seen the classification of a bat as 'yakt'. But who can say how the Karam would classify an animal hitherto unknown to them, an owl for example?[3]

In short, some sociologists of science have discerned a 'social' corridor between a paradigm and its application; a corridor which seems of particular importance when explaining the competition between paradigms and the outcome that leads to the predominance of one over the other. Put otherwise: 'why can one man's successful solution be another man's anomaly?' (Barnes, 1982b: 114).

Sociologists have given two types of answer to this question. The first imagines the 'corridor' as a narrow passageway substantially internal to the scientific community. According to this account, preference for a paradigm may be due to 'micropolitical' factors like the investments (in terms of effort, reputation, career) made by a particular group of researchers in a certain enterprise. For example, Pickering (1980) has studied the controversies of the 1960s and 1970s between physicist supporters of the 'charm' and 'colour' models of the quark. Although there were data supporting and contradicting both models, the majority of high energy physicists opted for the charm model because it enabled them to view quarks as real entities and not simply as models, and this set greater value on their work hitherto in studying quarks. Another example is the transition in French physics from the corpuscular to the undulatory paradigm of light in the early nineteenth century. This was not accomplished by converting the supporters of the former theory like Laplace and Poisson; rather, it resulted from the occupation of key positions in French science by an anti-Laplace faction (Frankel, 1976).

The second type of answer construes the corridor in terms of 'macropolitical' elements from the wider social context in which the scientific community in question operates. For example, a historical study by Forman (1971) relates the rise of quantum physics at the end of the 1920s to the critique waged against the determinism and concept of causality embraced by the intellectual elites of the Weimar Republic.

Of course, these two orders of factors may combine with each other. According to Wynne (1979), the reluctance of Cambridge physicists during the late Victorian age to abandon the concept of 'ether' can be analysed at various levels. Besides being of great usefulness in solving particular problems, while also preserving the scholar's previous work against dispute, the concept of ether sat perfectly with defence of the theological convictions and clerical institutions that characterized the aristocratic and land-owning class to which physicists belonged. The idea of 'ether' matched that of a harmonious cosmos and faith in a transcendent entity, and as such served to counter the instrumentalist doctrines of scientific naturalism (Barnes, 1982b).

An original solution to the problem of explaining the shift from one paradigm to another is provided by application of the anthropological 'grid/group' model (Douglas, 1970) to scientific communities. This model classifies social groups according to two features: the extent to which their internal relations are hierarchized and structured (grid), and their degree of cohesion vis-à-vis the outside (group). An army or a bureaucracy are examples of 'high grids'; a religious sect whose members have little contact with outsiders is an example of a 'high group'.

The upper right quadrant (high group, high grid) comprises more static scientific communities with strong internal cohesion and little competition. Here, an anomaly which threatens the paradigm will be ignored or rejected as 'freakish'; everything possible will be done to defend the current interpretative framework. In the lower left quadrant, where scant isolation from the outside combines with strong competition among members, an anomaly will be largely regarded as an opportunity. Scientists like Laplace and Poisson, firmly ensconced in the scientific establishment, were therefore less sensitive to the problems with corpuscular theory than were outsiders like Arago and Fresnel (Frankel, 1976).

This approach has been used to explain the 'methodological revolution' that swept through mathematics in around 1840. Hitherto, mathematicians had not given great importance to counter-examples and anomalies, nor did they use them dialectically to refine their proofs. Abel and Fourier found exceptions to Cauchy's hypothesis that the limit function of any convergent series of continuous functions was itself continuous; but between 1821 and 1847 the hypothesis and its exceptions lived tranquilly side by side, without mathematicians feeling obliged to reject Cauchy's hypothesis. According to Bloor (1982), the revolution – in which German mathematicians like Seidel played a leading role – was made possible by the reorganization of university research and teaching in Prussia. Centralization and the intervention of the government bureaucracy in professorial appointments broke up the self-centred circles and loyalties of academe, stimulating competition and inducing researchers to promote their careers with original discoveries.

These attempts to answer Barnes' question quoted above have had a twofold impact. On the one hand they have given greater importance to the social component along the vertical axis: political and social factors matter not only at the climax of scientific revolutions – as Kuhn theorized – but also at their beginnings and when they have subsided. In the early 1800s there was no crisis facing the proponents of corpuscular light; indeed, the theory was enjoying a period of great expansion and success. In 1897, heated controversy was provoked by Buchner's discovery that the cellular liquid extracted from yeast slurry fermented sugar into alcohol and carbon dioxide even in the absence of living cells. Was the fermentation caused – as Buchner and the exponents of nascent biochemistry maintained – by an enzyme (zymase)? Or was it due to residues of cellular protoplasm, as believed by the proponents of traditional cellular theory and technical experts on fermentation? Buchner's experiments lent themselves to different interpretations and convinced neither side; indeed, they were of service only to those who were already convinced. What they did do, however, was bring out a concrete and polarized debate – enzyme versus protoplasm – in which those who already subscribed to a biochemical view of cellular processes could mobilize and clarify the approach and goals of their sector, which was as yet uninstitutionalized. The change of paradigm – from a dichotomous protoplasm/ferment view to a unitary one based on enzymes – had to a large extent already come about, but it needed a focal point which would reveal 'like a prism ... the spectrum of existing attitudes toward vital phenomena' (Kohler, 1972: 351). Horizontally, the range of action of social and political factors extends to the wider social context.

One shortcoming of Kuhn's theory therefore resides in its tendency to consider – in order to explain scientific revolutions – solely the dynamics internal to the community of specialists. This is a weakness that does not seem to affect an author cited by Kuhn as one of his main sources of inspiration: Ludwik Fleck.

Fleck, a Polish doctor of Jewish origin, had published in 1935 an essay entitled *Genesis and Development of a Scientific Fact* (Fleck, 1935). Rediscovered and republished in numerous languages at the end of the 1970s, the text has become a classic in the sociology of scientific knowledge.

Fleck uses a practical example with which, as a doctor, he was well acquainted: the evolution of the concept of syphilis. As he follows the tortuous history of the concept, Fleck anticipates many of Kuhn's conclusions: each scientific fact acquires meaning within a particular 'thought style' – a term which he uses in more or less the same sense as Kuhn's 'paradigm'. Different conceptions of syphilis lead to the inclusion or exclusion under that pathology of cases which otherwise might be regarded as akin to chicken pox or other diseases. Unlike Kuhn, however, Fleck discovers that different 'thought collectives' (i.e. communities that share a certain 'thought style') 'intersect repeatedly in time and space'. Gravitating around a particular thought style are an esoteric circle (of specialists) and an exoteric circle of non-specialists. The thought style draws its strength from the constant interaction between these circles; in particular, it is the exoteric circle (i.e. at the 'popular' level) which displays thought styles in most clear-cut and incontrovertible manner. There may be doubts and fine distinctions, ambiguous observations and data among astrophysicists; but for the general public the 'Big Bang' is without question the origin of the universe. For physiologists there may be 'false positives', unclear patterns of bacteria under the microscope, HIV tests which give negative results even with patients classified as infected with AIDS; but for the public BSE is the prion disease, syphilis is the spirocheta pallida disease, and AIDS and HIV coincide (Berridge, 1992).

The researcher, as simultaneously the member of several thought collectives (the community of specialists to which s/he belongs, but also a political party, a social class, a culture), finds him/herself at the centre of these constant exchanges. Fleck shows that numerous themes in the modern conception of syphilis spring from collective ideas (what he calls 'protoideas'): the religious idea of 'disease as punishment for lust', or the ancient popular idea of 'syphilitic blood'.

According to Fleck, not taking account of this collective character of knowledge is like trying to explain a football game by analysing only the passes and moves made by the players one by one. Indeed, his conclusion is much more radical than that of many contemporary sociologists of science: 'Cognition is the most socially-conditioned activity of man, and knowledge is the paramount social creation' (Fleck, 1935, English trans. 1979: 42).

Notes

1 Of course, the labels used by Chia for the responses identified (e.g. positivism, constructivism) do not necessarily correspond to the epistemological positions denoted by the same names.
2 'Revolution' is a term of astronomical origin which means regular rotation; only latterly has it entered the political lexicon to signify radical change: see Cohen (1985).
3 What I have very briefly described here is what philosophers call 'finitism'. See Hesse (1966), Barnes (1982b).

3 Is mathematics socially shaped?
The 'strong programme'

1 The planet that could only be seen from France

The most important advance in nineteenth-century astronomy was the discovery of a new element in the solar system. Since 1781, when Laplace had hypothesized that this new element was a planet called Uranus, astronomers had observed deviations by the planet from its predicted orbit. In the early decades of the next century, a number of scientists suspected that these deviations might be due to another, hitherto undiscovered, planet. In 1845, a student at Cambridge, John Adams, calculated the orbit of this hypothetical planet and reported his findings to the Greenwich Observatory, which was nevertheless unable to detect it by telescope. In the meantime, the director of the Astronomical Observatory of Paris, Urban Jean Le Verrier, had independently reached the same conclusions and in 1846 announced the discovery of a new planet, to which the name of Neptune was given. The discovery was hailed as a triumph by the French scientific community, which used it as a watchword in its struggle against the Church for the monopoly of knowledge about nature. Then, however, the American astronomer Walker calculated a new orbit for Neptune which was entirely different from the one worked out by Adams and Le Verrier. Was this the orbit of the same planet or of a different one? For the American astronomers it was a different one; for the French astronomers, who had made massive investments in terms of their public image and scientific authority in Le Verrier's discovery, it could only be Neptune, and the different orbits could only be due to errors of calculation (Shapin, 1982).

The controversy over Neptune's orbit is typical of the cases examined by the tradition of science studies carried forward by the so-called 'Edinburgh School'. After its foundation in 1966 by the astronomer David Edge, the Science Studies Unit of Edinburgh moved rapidly to the forefront in the social studies of science. Since then, Barry Barnes, David Bloor, Donald MacKenzie, Steven Shapin and Andrew Pickering are some of the scholars who have worked at the Unit. When first developing their approach to the sociology of science, the firm intention of these scholars was to oppose the institutional sociology of science that had become established in the US since the Second World War. The punctilious definition given to their subject of study as the 'sociology of scientific knowledge' (SSK), rather than simply as 'sociology of science', was an explicit declaration of intent to open the 'black box' of science which, in the opinion of the Unit's members, the institutionalized approach had left largely intact, doing no more than examine its external features.

Whereas the approach of Merton and his followers belonged largely within the sociological mainstream, the approach of the Edinburgh School has been clearly interdisciplinary from the outset. It makes extensive use of materials from the history of science (as well as conducting original case studies, although almost always from a historical perspective) and it engages in constant dialogue – albeit often critically – with the philosophy of science.

It should be emphasized that the SSK theorized at Edinburgh is based on case studies, and that it has simultaneously stimulated a large body of work by sociologists and historians of science. A valuable essay by Steven Shapin has organized this mass of studies into four broad areas on the basis of the analytical aims and significance of each of them.

The first area comprises studies that highlight the *contingent* nature of the production and evaluation of scientific findings. In other words, these are studies which reveal the existence of a 'grey area' between what nature offers to researchers and their accounts of it, and that this grey area may, in principle, comprise factors of a social nature.

For example, in 1860 the English biologist T.H. Huxley announced the discovery of a primitive form of protoplasm which he called *Bathybius Haeckelii*. His discovery was soon confirmed by other scholars, and the *Bathybius* was, for a long time, considered to be a 'fact', being cited in support of the nebular hypothesis of planetary evolution by numerous Darwinians, as well as by Huxley and Haeckel themselves. The *Bathybius* was taken to constitute proof of the continuity between non-living forms and living beings. Only subsequently did certain biologists begin to argue that the *Bathybius* was an artefact bred from a combination of 'observers' imagination and the precipitating effect of alcohol on ooze' (Shapin, 1982: 160).

Is mathematics socially shaped? 43

In entirely similar manner, cellular meiosis was observed or denied by various groups of researchers until – following 'rediscovery' of Mendel's theories in the early twentieth century – chromosomic theory came up with an interpretative grid able to accommodate cytological observations. Golgi's corpuscle is another fact/artefact that has long made cyclical appearances and disappearances in observations by cellular biologists (Dröscher, 1998).

Shapin himself, however, admits that these studies

> open the way to a sociology of scientific knowledge [but] they do not by themselves constitute such a sociology. An empirical sociology of knowledge has to do more than demonstrate the *underdetermination* of scientific accounts and judgements; it has to go on to show *why* particular accounts were produced... and it has to do this by displaying the historically contingent connections between knowledge and the concerns of various social groups in their intellectual and social settings.
>
> (Shapin, 1982: 164, my italics)

This goal is achieved, according to Shapin, by the studies belonging to the second area – the one which uses professional interests as an element in sociological explanation. In the already cited case of the *Gilia inconspicua* (see Chapter 2), the criteria used by both sides to argue for the superiority of its own classification of the plant can be related to the desire of each to protect its conspicuous investments in learning, publications and reputation. The hypothesis that there exist tumour-provoking viruses – which subsequently won Temin, Baltimore and Dulbecco the Nobel prize for their discovery of the reverse transcriptase enzyme – inevitably provoked the scepticism of scientists who had spent lifetimes working under the 'dogma' that RNA could never generate DNA (Kevles, 1999). It is not rare for such conflicts to arise among scientists of different scientific affiliations. English biologists, unlike geologists, had been inclined to abandon a teleological view of natural history already before publication of Darwin's *Origin of the Species* (1859).

A theory that the adaptation of living beings was governed by biological laws, and not by a divine plan or by simple environmental determinism, enabled biology to free itself from the sway of geology; for geologists, by contrast, a teleological account enabled them to treat geological change as primary and that of living beings as its consequence (Ospovat, 1978, cf. Shapin, 1982). When the dispute erupted over the alleged discovery of cold fusion by Pons and

Fleischmann in 1989, chemists and physicists were not only in conflict over their respective purviews (who should study the phenomenon) but also over which signals constituted 'proof' that fusion had occurred: the production of heat according to the chemists, the emission of neutrons according to the physicists (Lewenstein, 1992a; Bucchi, 1996). During the already-mentioned controversy over zymase,[1] industrial mycologists were uninterested in detailed analysis of the cell's inner functions, which were of little relevance to their work; while those who had publicly supported the protoplasm theory were strenuously opposed to any recognition at all of zymase. From a theoretical point of view, the new results could be reinterpreted in the light of the old protoplasm theory, adapted so that a role could be given to enzymes. Yet, in the social domain the debate had by now polarized between two irreconcilable camps, with zymase being brandished by the biochemists as the symbol of a new era and of the struggle against the old establishment (Kohler, 1972).

According to SSK, what scientists 'see' and the explanations they give for it relate more generally to the role of science and scientists at a given historical moment, and to the level of professionalization and separation between experts and non-experts. This is the theme of the third area of studies singled out by Shapin. In the seventeenth century, French academics were reluctant to accept that meteorites came from the sky because accounts of their fall very often originated from peasants, or at any rate from 'non-professionals'. They were consequently deemed unreliable. Following the Revolution and, consequently, the change in attitude among intellectuals towards the common people, scientists began seriously to consider the connection between meteor showers and the fall of rocky objects in the countryside.

The fourth group of studies cited by Shapin enable him to argue that the role of social factors does not stop when scientific activity has been professionalized. In fact, it is possible to show that scientists make much use of images, models and metaphors from the more general culture at large. The source of these images may be for instance technological (an example being the mechanical pumps to which Harvey compared the heart) or political culture. The great biologist and political activist Virchow, for example, presented his conception of the organism made up of cells through analogy with his solidarist conception of a society in which individual citizens cooperate in the collective interest (Mazzolini, 1988). Better known and more widely studied is the influence exerted by Malthus' theory of social competition and individualism – ideas which pervaded Victorian society – on Darwin's development of his evolutionary theory (Gale, 1972; Young, 1973). George Poulett Scrope, one of the first geologists to hypothesize constant and long-period geological processes – thereby helping to discredit 'diluvial' explanations – also studied and wrote about political economy. His use in geology of the concept of time as an explanatory factor – 'neutral' with respect to other events, and potentially infinite – derived from his view of money as a means of circulation and exchange bereft of any intrinsic value (Rudwick, 1974).

Evelyn Fox Keller (1995) has described the history of biology in the twentieth century as the shift between two paradigmatic 'metaphors': a transition, that is, from a metaphor centred on the embryo and the organism's gradual development to one attributing to the gene – equivalent to the atom in physics – the capacity to 'construct' the organism on a predefined template. The former metaphor has been dominated by embryology; the latter has been characterized by the rise to predominance of genetics. This transition can be interpreted at various levels. One of them is specifically technical and has radically transformed the conditions and potential of biological research; the other is political and concerns the opposition and subsequent reconciliation between Germany – where the embryological paradigm held sway – and the US, where the genetic paradigm rapidly rose to dominance. At the cultural level, the genetic paradigm owes a great deal to the concept of information developed in cybernetics. And at an even broader cultural level, the waning of genetic determinism and the rediscovered importance of the 'cytoplasm' – the female part of the cell – owe a great deal to the feminist movement of the second half of the twentieth century.

The process also operates in reverse: images and concepts from science may be transferred into the political and social spheres. According to the SSK approach, the theories or explanations selected for such transfer depend on the specific circumstances of certain social groups, and on the specific strategies pursued by them.

An example is provided by phrenology. Developed during the nineteenth century from the work of the German doctor Franz Joseph Gall, this doctrine maintained that a person's psychological characteristics are located in specific zones of the brain, to which correspond bumps on the cranium. In the years around 1820, the theory provoked heated debate at Edinburgh University between phrenologists and anatomy lecturers. The dispute centred on different conceptions of the brain. This the university anatomists viewed as a unitary whole, whereas the phrenologists believed that it was an assembly of parts corresponding to different intellectual faculties. Both groups were made up of distinguished anatomists, and both groups performed careful dissections and examinations of the brain. For Shapin, phrenology gave the mercantile class the ideal means with which to challenge the academic elites. By turning phrenology into a dynamic theory of heredity, they could use it to highlight, besides the existence of certain traits inherent to the individual, also the possibility of altering or changing those traits by means of social reform. Not coincidentally, this view of heredity grew more entrenched as the bourgeoisie found itself having to cope more and more with the working class's demands for reform, and shifted its favour to eugenic theories in consequence (MacKenzie, 1976).

Thus, what Shapin calls *full circle* is achieved: 'connecting interests in the wider society to judgements of the adequacy and validity of esoteric mathematical formulations' (Shapin, 1982: 191). It is wrong, Shapin maintains, to yield to the temptation of separating the strictly technical component of a controversy from its 'cosmopolitan and methodological' ones.

> Anti-phrenologists' insistence that cranial bones in the region of the frontal sinuses were not parallel was explicitly connected to their claim that phrenological character diagnosis was impossible; phrenologists' assertion that the cerebral convolutions might show standard pattern and morphological differentiation was explicitly related to their view that mental faculties were subserved by distinct cerebral areas.
>
> (Shapin, 1982: 193–194)

We may likewise read the controversy on heredity that broke out in the early twentieth century between the biometrics school and the Mendelians. While the former propounded a rigid Darwinism, whereby evolution was the constant selection of minuscule differences, the latter embraced Mendel's recently rediscovered theories and their underlying hypothesis of more abrupt and discontinuous changes. According to Barnes and MacKenzie, this contrast reflected not only different technical competences and resources – for example, the biometricians made much use of mathematical-statistical tools – but also more general political and social attitudes. The biometric approach was compatible with the eugenic convictions and social reformism of the middle class, which pressed for political measures capable of shaping the development of society. The Mendelian approach instead reflected the conservative and non-interventionist views of the more reactionary classes (Barnes and MacKenzie, 1979).

Is mathematics socially shaped? 47

These dynamics have also been used to analyse the controversy in statistics between Pearson – the leader of the biometrics school – and Yule. The dispute centred on the most appropriate correlation indicator for nominal statistical variables like 'living/dead' or 'high/low'. The index proposed by Pearson – rt – was based on the hypothesis that such variables can be considered products of a bivariate normal distribution. Yule instead developed another index – Q – which dispensed with that assumption. In this case, too, the incompatible positions taken up (and backed by opposing 'networks' in the British academic community) can be linked with the different goals that Pearson and Yule believed that statistical theory should pursue. What was assumed to be 'normality', however, depended on the scientist's broader vision of society – which in Pearson's case was centred on eugenics and Fabian socialism (MacKenzie, 1978).

A further example is provided by the history of Italian mathematics and concerns one of the last of Italy's mathematical 'duels', which was held in Naples in 1839. The tradition of mathematical duels dated back to the Renaissance, when they were frequently used to settle scholarly disputes. Originally watched by a crowd of spectators as two or more mathematicians strove to solve the same problems, with time these duels came to be conducted by correspondence or in the columns of learned journals. The duel in Naples resulted from a challenge issued by the mathematician Vincenzo Flauti against members of the 'analytic' school, whom he invited to solve three problems of geometry. A professor at the University of Naples and secretary to the Royal Academy of Science, Flauti was the leading exponent of the 'synthetic' school, whose teaching centred on pure geometry and the methods of classical mathematics. The founder of the school, Vincenzo Fergola, a fervent Catholic and the author *inter alia* of essays which asserted the effectively miraculous nature of the liquefaction of Saint Januarius' blood, considered mathematics to be a 'spiritual science', on the grounds that it was pure, and consequently insisted that it should not be contaminated with practical applications. The analytic school was institutionally associated with the *Scuola di Applicazione del Corpo di Ingegneri di Ponti e Strade*, which trained bridge and road engineers, and was therefore more concerned with geometrical analysis and the application of calculus to empirical problems. The two schools had been at loggerheads since the beginning of the century, with the 'analyticists' accusing the 'syntheticists' of anti-scientific behaviour because they had ignored the algebraic revolution in French mathematics; while the syntheticists responded in kind, going even so far as to accuse their rivals of moral depravity.

48 Is mathematics socially shaped?

In the end, the mathematics section of the Royal Academy, which was given the task of adjudicating the duel and awarding a monetary prize to the winner, pronounced against the analytic school: a judgement prompted, according to several scholars, by the closer compatibility of the synthetic school with the counter-revolutionary policy of the Bourbons and the Catholic Church (Mazzotti, 1998).

What conclusions can we draw from these various examples? Shapin warns against adopting the unsatisfactory and caricatured version of the sociology of knowledge which he calls the 'coercive model'. This model, in fact:

a   claims that sociology asserts that all individuals in a certain social situation will adopt a certain intellectual belief;
b   treats the social as a mere aggregation of individuals;
c   establishes a deterministic relationship between social situation and beliefs;
d   views sociological explanation as concerned with 'external' macrosociological factors;
e   opposes sociological explanation to the assertion that scientific knowledge is empirically grounded on sensory inputs from natural reality.

None of these statements reflects the SSK approach and its thesis that 'people produce knowledge against the background of their culture's inherited knowledge, their collectively situated purposes, and the information they receive from natural reality'. In this regard, the exponents of the SSK have taken especial pains – and here again they depart sharply from the Mertonian tradition – to reconstruct in detail the activities, methods and concrete experimental practices of scientists. Many of the members of the Science Studies Unit, moreover, had scientific backgrounds: Edge came to it from astronomy, Barnes from physics and Bloor from cognitive science. 'The role of the social' concludes Shapin 'is to prestructure [scientist's] choice, not to preclude choice' (Shapin, 1982: 196, 198).

2 Is even mathematics 'social'?

The proponents of the SSK have examined the relationship between science and society from various points of view. Yet the Edinburgh school has often been identified – by its critics especially – with the so-called 'strong programme', the classic formulation of which was set out by David Bloor in his *Knowledge and Social Imagery*

(1976). Although Bloor and his book have been regarded – again by critics especially – as epitomizing the sociology of science, it should be borne in mind that Bloor developed his interest in the philosophical and sociological analysis of science after earning a doctorate in psychology. His main intention, as he recalls today, 'was to show to philosophers of science that in the light of a wide range of studies, mainly carried out in the history of science, it was not possible anymore to hold a vision of science as exempt from social influences'.[2]

The core of the 'strong programme' consists of a set of methodological principles for the sociological analysis of scientific knowledge. According to Bloor, such analysis should be:

(i) *Causal*, i.e. concerned with the conditions which bring about beliefs or states of knowledge.
(ii) *Impartial* with respect to truth or falsity, rationality or irrationality, success or failure. Both sides of these dichotomies require explanation.
(iii) *Symmetrical* in its explanation. The same types of cause should explain true beliefs and false ones.
(iv) *Reflexive*. In principle its patterns of explanation should be applicable to sociology itself, which obviously cannot claim to be exempt from sociological analysis.

(Bloor, 1976: 4–5)

Bloor obviously does not deny that there exist 'other types of causes apart from social ones which will cooperate in bringing about belief', but his intention is to give greater dignity and pervasiveness to sociological explanation. Social factors like interests, political ideologies and cultural features, he maintains, should not be brought to bear solely when knowledge jumps the rails of rationality or lapses into error. This attitude – which Bloor views as characterizing most of the preconceived objections made against the sociological approach to the study of science – sees 'logic, rationality and truth' as 'their own explanation ... it makes successful and conventional activity appear self-explanatory and self-propelling' (Bloor, 1976: 6). On this view, sociological explanation should only intervene when some anomaly (which cannot but be 'social') deviates rationality and progress towards the truth from their automatic course. Sociology could thus explain – by invoking religious or political or more generally cultural factors – Kepler's mystical beliefs about the sun, or the astronomer Schiaparelli's conviction that Mars was populated by human beings organized into some sort of socialist collective. It could also explain the 'Lysenko case' – that of the Stalinist biologist who for many decades suppressed the Mendelian theory of genetically transmitted traits, arguing in obeisance to communist ideology that they instead depended on environmental conditions. But it could not explain the factors responsible for the success of Darwinism or of Virchow's cellular theory. It is this 'weak programme' that Bloor's theoretical proposal opposes.

In order to illustrate the symmetry principle, Bloor refers to a comparison made by Morell between two schools of chemistry research in the early 1800s: Liebig's school at Giessen, and Thomson's school in Glasgow. According to Bloor, the radically different fortunes of these two schools (international success for Liebig's, oblivion for Thomson's) cannot be explained solely on the basis of the experimental results achieved by the two great scientists. Also responsible were factors such as the personalities of the scientists who headed the schools; their status and relative abilities to obtain funding for their laboratories; and their choice of sector in which to conduct their research. For example, Thomson was working in a political context where it was impossible to obtain public funding, which was instead amply available to Liebig. In his dealings with his pupils, Thomson tended more to exploit their labour than to set value on it. Finally, Thomson chose to work in a mature sector, that of inorganic chemistry, where experts like Berzelius and Gay-Lussac had already made glittering reputations, and where it was difficult to come up with innovative and significant results. The sector of organic chemistry chosen by Liebig was of more recent development, less structured and less dominated by other researchers, and it was characterized by simpler experimental procedures, easier to teach to pupils.

A possible objection against the strong programme is the so-called 'argument from empiricism', which runs as follows: 'social influences produce distortions in our beliefs whilst the uninhibited use of our faculties of perception and our sensory-motor apparatus produce true beliefs' (Bloor, 1976: 10). Bloor meets this objection by pointing out that an increasingly negligible part of knowledge – and scientific knowledge in particular – derives directly from the senses. The perception of scientists themselves – not to speak of non-scientists – is mediated by complex instruments and by elaborate intermediation apparatus (publications, experimental equipment, the mass media).

It is therefore impossible to distinguish sharply between 'truth = individual experience' and 'error = social influence'. Indeed, it is precisely the social dimension (the sharing of standardized experimental practices, agreement on criteria and procedures, repeatability and controls) that guarantees the functioning of science despite distortions in the individual perceptions of researchers. It is not brute experience or observation that stands at the centre of scientific activity but socialized activity, 'repeatable, public and impersonal' (Bloor, 1976: 26).[3]

To illustrate the point more thoroughly, Bloor recounts the well-known story of Blondlot's N-rays. Blondlot, a French physicist and member of the Academy of Science, announced in 1903 that he had discovered a new type of radiation similar to X-rays. One of the properties of his N-rays was that they were polychromatic: when passed through an aluminium prism, Blondlot claimed, they could be shown to comprise elements with different indices of refraction. During a visit to Blondlot's laboratory, the American physicist Robert Wood surreptitiously removed the prism; even so, Blondlot continued to see signals emitted by the N-rays. Wood wrote an article about his visit for the journal *Nature* in which he concluded that N-rays did not exist: they had simply been produced by Blondlot's desire to discover another type of radiation.[4]

'Sociologists', Bloor comments,

> would be walking into a trap if they accumulated cases like Blondlot's and made them the centre of their vision of science. They would be underestimating the reliability and repeatability of its empirical base; it would be to remember only the beginning of the Blondlot story and to forget how and why it ended. The sociologist would be putting himself where his critics would, no doubt, like to see him – lurking amongst the discarded refuse in science's back yard.
>
> (ibid.: 25)

The point for Bloor is not that observation or data from experience are valueless; rather, the point is that they do not suffice in themselves to bring about change in beliefs. Bloor depicts the relationship between experience and beliefs as in Figure 3.1.

Scientific theories and results are often 'under-determined' by observational data. In this regard Bloor furnishes a series of examples of how the same perceptive or observed data can be interpreted in completely different ways. He cites the elementary case of the apparent diurnal movement of the sun, which has been interpreted in different epochs and observational contexts as demonstrating the sun's rotation around the earth, but also the other way round. Another example is the 'parallel roads' along the sides of Scottish valleys;

52 Is mathematics socially shaped?

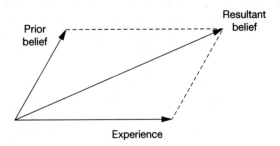

Figure 3.1 The relationship between experience and beliefs
Source: Bloor (1976: 27)

these are geological phenomena even though they look like manmade paths. On the basis of his observations of similar 'roads' during his travels in South America, Darwin thought that they were due to the erosive action of the sea; Agassiz, a geologist who had studied the Swiss glaciers, offered the entirely different explanation that they resulted from lakes imprisoned during the Ice Age.

The geologist Alexander Du Toit – among the first to endorse Wegener's hypothesis of continental drift when it was still being dismissed as absurd by a large part of the scientific community – lived in South Africa, and there the evidence of the break-up of the continents was more obvious than elsewhere. His contribution to the theory was to replace Wegener's Pangaea with two original continents, Luarasia and Gondwana, with the centre of the latter located precisely in what is today's South Africa.

Whereas Priestley, on placing a gas flask in a water bath on which a small pot of minium was being heated, saw the red lead absorb phlogiston and change into lead, we, today, see the oxygen separate itself from the lead oxide and leave the lead as a deposit.

Bloor even goes so far as to apply the strong programme to the scientific discipline usually considered most impermeable to the influence of social factors: mathematics. His concern in this case is to show that even formulas, proofs and elementary results do not have an intrinsic meaning but depend on a set of presuppositions. The proof that the square root of two is an irrational number may lose significance in a mathematical system in which the concept of even and odd do not exist; or it may be interpreted (as it was by the Greek mathematicians) as proof that the square root of two is not a number at all. To different institutional and cultural contexts may correspond

Is mathematics socially shaped? 53

different logics or mathematics. Even the solution of a mathematical problem may be the result of a complex negotiation. In this regard, Bloor takes from Lakatos (1976) the example of Euler's well-known theorem on polyhedra which relates their number of vertices, edges and faces thus:

$$V - E + F = 2$$

To this theorem, which was formulated inductively by Euler in 1752 and demonstrated by Cauchy in 1813, Lhuiler and Hessel found an exception: the polyhedron shown in Figure 3.2, which satisfied the standard definition (a solid whose faces are polygons) but not Euler's theorem. It was therefore necessary to reformulate the definition of a polyhedron as a 'surface composed of polygonal faces'. Shortly afterwards, however, further exceptions were discovered, like that shown in Figure 3.3. This time it was the proof that had to be reformulated as being valid only in the case of simple polyhedra – ones, that is, whose faces could be flattened. But Figure 3.4 shows a simple polyhedron for which Euler's theorem does not hold.

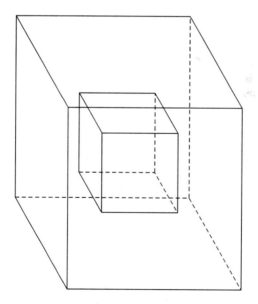

Figure 3.2 Lhuiler and Hessel's polyhedron
Source: Bloor (1976: 133)

54 *Is mathematics socially shaped?*

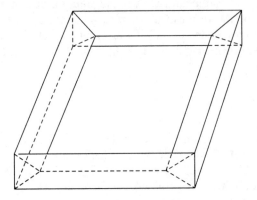

Figure 3.3 An example leading to the reformulation of Euler's theorem
Source: Bloor (1976: 134)

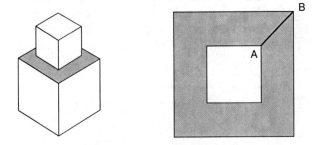

Figure 3.4 A further exception to Euler's theorem
Source: Bloor (1976: 135)

According to Bloor, this example shows that not even in mathematics are there immutable definitions and postulates from which proofs and theorems 'automatically' and invariably derive. Instead, constant negotiations take place over the definitions themselves; negotiations which, in the specific case of Euler's theorem, concerned what a polyhedron actually is and whether exceptions should be incorporated into the theorem by modifying it, whether they should be rejected as 'non-polyhedra' (perhaps by restricting the definition), or whether they should be deemed to confute the theorem. The choice of one or other of these options can be related to the social and institutional context in which the researcher is working. For example, a closed and strongly cohesive scientific community based on loyalty to a specific theory or result, and where greatest value is set on obedience to tradition, may see any counter-example as a threat to its existence, and therefore tend to expel exceptions to the Euler/Cauchy theorem, calling them – as Mathiessen did, for example – 'recalcitrant cases' (cited in Bloor, 1982: 200). In a more differentiated context, where diverse groups of mathematicians work in diverse institutional settings (academies, universities, journals), an anomaly can live together with the rule: the theorem can be retained with certain restrictions or deemed valid under certain conditions; 'no formula has indeterminate validity', was Cauchy's riposte to the counter-examples brought against his theorem. Finally, a highly competitive and individualistic context, in which originality and innovation are rewarded, will opt for a 'revolutionary' response and therefore abandon the theorem (see Chapter 2).

3 The weaknesses of the strong programme

Though generally recognized as ambitious, Bloor's endeavour has been considered by several critics as not entirely successful. Some of them have argued that if the declared objective of his work, and that of the Edinburgh school in general, has been to delve into the 'black box' of science – at whose exterior Merton came to a halt – it has not been completely achieved.

Perhaps with excessive over-simplification, a philosopher of science particularly critical of the sociological approach has singled out four versions of what he calls 'externalism' (the view that the context is able to determine the content of scientific research) (Bunge, 1991):

(a) Moderate or weak externalism: knowledge is socially conditioned.
(a1, local) The scientific community influences the work of its members.
(a2, global) Society as a whole influences the work of individual scientists.

(b) Radical or strong externalism: knowledge is social.
(b1, local) The scientific community constructs scientific ideas.
(b2, global) Society as a whole constructs scientific ideas.

Bloor's approach seems, at times, to restrict itself to conceptions little different from those of Merton and his school, lying midway between (a1) and (a2) – especially when it analyses the influence of factors like the style of the leaders of the Liebig and Thomson schools, and more generally of the economic-social context, on their differing fortunes. Elsewhere, Bloor appears to adopt a perspective close to Kuhn's, or especially Fleck's, when he argues that it is theoretical predispositions or proto-ideas that guide observation or the conduct of experiments, not the other way round.

It is not that these various gradations are mutually incompatible. Indeed, Bloor sometimes seems to theorize a kind of sociological opportunism whereby the role of the social component may vary from a minimum to a maximum according to the type of scientific case under examination. 'When the signal noise ratio is as unfavourable as this' – the reference being to Blondlot, but also to Huxley and his *Bathybius* or Golgi's corpuscle – 'then subjective experience is at the mercy of expectation and hope' (Bloor, 1976: 25).

But the danger of this attitude is that it may push sociology back into the residual role of dealing with the 'rejects' of science (gross errors, cases of deviance) – a role which Bloor explicitly opposed, and to do so formulated the symmetry principle.

Numerous critics have pointed out the ambiguity of this principle. According to Ben-David, for instance, the examples furnished by Bloor do not satisfy the criteria of covariance and causality. If a specific interest or cultural orientation determines the adoption of a particular scientific perspective, then a change in the former should necessarily give rise to a change in the latter. But this obviously does not always happen: numerous theories or approaches may succeed one another in the same political or cultural context. Bloor responds to this objection by restating the claims of his approach: '[This point] would be fatal only to the claim that knowledge depends *exclusively* on social variables such as interests' (Bloor, 1991: 166, italics in the original).

> Doesn't the strong programme say that knowledge is purely social? ... No. The strong programme says that the social component is always present and always constitutive of knowledge. It does not say that it is the *only* component, or that it is the component that must necessarily be located as the trigger of any and every change: it can be a background condition. Apparent exceptions of covariance and causality may be merely the result of the operation of other natural causes apart from social ones.
> (Bloor, 1991: 166, italics in the original)

A more sophisticated criticism has been brought against the relationship between social and 'natural' factors. Consider again the example of phrenology. According to Shapin, the two sides in the controversy differed in their views because they came from different social backgrounds. While the anatomy lecturers were an elite characterized by an esoteric notion of knowledge, most of the phrenologists were amateur scientists, often tradesmen or members of the middle class, who espoused a more 'accessible' conception of science.

The objection by scholars like Brown is that the under-determination of theories with respect to data does not automatically entail that interests play a decisive role. 'In fact,' Brown objects, 'just as there are infinitely many different theories which will do equal justice to any finite set of empirical data, so also are there infinitely many theories which will do equal justice to a scientist's interests' (Brown, 1989: 55).

In other words, if it was the intention of the Edinburgh middle classes to undermine the cultural hegemony of the aristocracy, why did they choose precisely phrenology for the purpose? Was the synthetic school in Naples the only mathematical approach compatible with the political and religious concerns of the Bourbon and religious authorities? What is it that makes social factors and scientific theories overlap?

Bloor's answer is plausible, as he says that there was no necessary reason for the opponents of the university elite to choose phrenology rather than any other theory for their purposes. 'Perhaps anything materialistic, empiricist and non-esoteric would have served as the not-X to the elite X' (Bloor, 1991: 172). 'Once chance favours one of the many possible candidates,' concludes Bloor, 'it can rapidly become the favoured vehicle', thus flanking the causality principle with a randomness principle. I shall return to this point later, because I believe it to be of considerable importance, though perhaps in a sense slightly different from that envisaged by Bloor.

Another weakness pointed out in the approach is its tendency to identify the social with interests, even though its proponents often use the latter term in a broader sense than mere material interests. The linkage between the cognitive dimension (the interpretative flexibility of which Bloor provides numerous examples at the micro-sociological level) and the macrosociological one of interests and social circumstances have sometimes been regarded as not made fully explicit. Paradoxically, two opposite critical reactions have been put forward on this point. On the one hand, the Edinburgh authors have been accused of transforming scientists into 'interest dopes'[5] or 'flat, puppet-like characters who were shaped by exogenous interests rather than a complex set of contingencies and motivations' (Hess, 1997: 92). On the other hand, it is possible to discern at the basis of the strong programme an equally idealized image of the omniscient scientific actor, perfectly rational, and able to choose consciously between one theory or method and another on the basis of his or her interests and those of the group to which s/he belongs.

What is certain is that the charge of radical relativism and constructivism brought against Bloor is largely unwarranted. And not only because he himself considers his mission to have been a 'positivist' attempt to apply a scientific method to the study of the relationship between science and society.[6] This is borne out by the consideration that over the years the strong programme has also been subjected to fierce 'internal' criticisms by sociologists of science themselves, and with regard to two aspects in particular. The first is the just-discussed one of causality. The SSK, the argument runs, does not greatly differ from Merton's model and that of the institutional sociology of science, for it does no more than replace norms with interests as the factors explaining how scientists behave. A large part of the studies discussed in the following chapters have been prompted by the more or less explicit intent to find alternatives to Bloor's allegedly too rigid model.

The second set of criticisms centres on the final 'commandment' of the strong programme: reflexivity. Some sociologists of science have emphasized the scant ability of the SSK theorists to apply the tools developed by themselves to the sociological analysis of scientific knowledge. The alternative proposed is that new narrative forms – dialogue, multi-voice or first-person narrative – should be used to bring out the nature as constructs of their own texts (Woolgar, 1988) or to make the 'social positioning' of their own observations explicit, as has been later attempted by feminist strands of science studies (Haraway, 1997).

Notes

1   See Chapter 2.
2   Personal communication, 4 June 1999.
3   Studies like those by Shapin and Schaffer on the controversy between Hobbes and Boyle have shown in more detail how the adoption of the 'empirical style' by science results from a complex historical-social process (Shapin and Schaffer, 1985). Today known only for his political theories, in seventeenth-century England Thomas Hobbes was also an active proponent of natural philosophy. His search for stability in natural philosophy based on logical argument, and according to which the very concept of vacuum was to be repudiated, found rebuttal by Boyle with an instrument that settled the matter: a machine able to 'produce facts',

namely the air pump used in his experiments on the vacuum at the Royal Society. A 'local' experiment witnessed by a restricted number of gentlemen – and who were therefore trustworthy – and then written up in detail was transformed into the 'matter of fact' able to bring everyone to agreement (see also Chapter 7).

4   Ashmore (1993) has analysed Wood's report in detail, showing that a 'trick' – surreptitiously removing Blondlot's prism – non-repeatable and more of an experiment in social psychology than physics, has been unproblematically incorporated into the literature and celebrated as epitomizing the scientific method, even by philosophers and sociologists of science.

5   The expression is used by analogy with that of 'cultural dope' coined by the founder of ethnomethodology, Harold Garfinkel, with reference to the way in which traditional sociological theories, especially Parsons', view the individual (Garfinkel, 1967).

6   Personal communication, 4 June 1999.

4 Inside the laboratory

1 A fascinating experiment

Likely hypotheses have been put forward on the trigeminal (Loewenstein *et al.*, 1930), bitrigeminal (Von Aitick, 1940), quadritrigeminal (Van der Deder, 1950), supra-, infra- and inter-trigeminal (Mason & Ragoun, 1960) afferents, as well as on the macular (Zakouski, 1954), saccular (Bortsch, 1955), utricular (Malosol, 1956), ventricular (Tarama, 1957), monocular (Zubrowska, 1958), binocular (Chachlik, 1959–1960), triocular (Strogonoff, 1960), auditive (Balalaika, 1515) and digestive (Alka-Seltzer, 1815) inputs.

On first reading this passage, it seems to be an extract from a *bona fide* scientific article. But as one reads the text that follows, the realization dawns that the aim of the experiment described was to determine the effect of tomato throwing on the voice volume of sopranos. The article is thus evidently a parody, the work of the French writer Georges Perec (Perec, 1991). But the fact that it is possible to write a parody of this kind, and that the reader may find it humorous, demonstrates that the scientific article – the so-called 'paper' – is by now a well-established genre of text and discourse, with precise codes and expressive rules as regards the abstract, the graphs, the tables, the acknowledgements, and so on.

The process involved in construction of a scientific article on the basis of informal conversations in the laboratory, experimental trial and error, ad hoc adjustments of hypotheses and explanations, has been examined since the mid-1970s by a series of studies which have sought to resolve the difficulties of the 'strong programme'. Such studies no longer take a certain scientific theory and set it in relation to a specific historical and social context; rather, they delve into the process itself that leads to the theory's formation, isolating its components and placing them under a magnifying glass.

This distancing from a certain naturalism and positivism – however paradoxical it may seem – apparent in the Edinburgh school, and in the strong programme in particular, has merged since the 1970s with stimuli from certain currents of sociological inquiry – notably ethnomethodology.[1] The founder of ethnomethodology himself, Harold Garfinkel, published an article in 1981 in which he analysed the discovery of a pulsar by a group of American astrophysicists, using for the purpose recordings that he had made of their conversations while they performed their observations and measurements (Garfinkel *et al.*, 1981).

This new approach therefore flanks the macrosociological and causal analysis of the strong programme with detailed inquiry into the contingent processes that constitute scientific activity. The method does not consist of attempts at systematic theory-making *à la* Bloor but, rather, of case studies whose minute reconstruction is often so complex that it takes up an entire book. The scientific fact is no longer seen as the point of departure; it is now the point of arrival. Scientific knowledge is not only socially conditioned – that is, social forces enter the internal procedures of science at a certain stage – instead, it is from the very beginning 'constructed and constituted through microsocial phenomena' (Latour and Woolgar, 1979: 236).

Unlike in the strong programme, analysis does not deal with historical cases but concentrates instead on contemporary science. The main setting for this microsociological and ethnographic observation is, therefore, the laboratory. In *Laboratory Life*, the first classic in this strand of studies, Latour and Woolgar (1979) spent two years observing the work of a research group at the Salk Institute of La Jolla, California – work which later led to discovery of a substance called TRF which earned Guillemin the Nobel prize. Latour and Woolgar analysed laboratory notebooks, experimental protocols, provisional reports and drafts of scientific papers, while carefully recording the conversations that went on during experiments and among the members of the research group. What were the conclusions of this and similar studies? According to another proponent of this approach, laboratory studies have shown that there are no significant differences between the search for knowledge that takes place in a laboratory and what happens, for example, in a law court. In scientific research, too, everything is, in principle, negotiable: 'what is a microglia cell and what is an artefact, who is a good scientist and what is an appropriate method, whether one measurement is sufficient or whether one needs to have several replications' (Knorr-Cetina, 1995: 152).

Involved in these negotiations are not only scientists but also the agencies that finance them, the suppliers of apparatus and materials, and policy makers, so that some scholars have been prompted to talk of 'transepistemic' networks. The across-the-board nature of these negotiations and the 'decision-impregnated' character (active, therefore, rather than being the passive recording of natural phenomena) of scientific research entail, according to Knorr-Cetina, the use by researchers of 'nonepistemic arguments' and their 'continuously crisscrossing the border between considerations that are in their view "scientific" and "nonscientific"' (Knorr-Cetina, 1995: 154). Playing a significant part in the construction of a scientific fact is the rhetorical dimension: discourse strategies, representation techniques, forms of data presentation. In this respect, Latour and Woolgar give particular importance to two groups of rhetorical items: 'modalities' and 'literary inscriptions' (Latour and Woolgar, 1979). Modalities are the elements that qualify the researcher's statements and which are gradually eliminated as a set of assertions or results is transformed into a scientific fact.

A sentence in a paper given to a seminar or a conference:

> The research group headed by Prof. So-and-So believes that there is some probability that beta-carotene may be involved in the prevention of some types of tumour.

In a textbook, or even more so a news magazine, this sentence will be transformed into:

> Beta-carotene prevents cancer.

Inscriptions are the 'evidence' – tables, graphs, microscope images, X-rays – that the researcher cites in support of his/her claims, almost as if to say, 'You doubt what I wrote? Let me show you' (Latour, 1987: 64). For Latour and Woolgar, therefore, a scientific instrument is nothing but an 'inscription device', an item of apparatus – whatever its technical sophistication, cost or size – able to produce 'a visual representation in a scientific text'.

The final outcome of this process is the article published in a scientific journal, where the researcher's progressive adjustments and zig-zag path are straightened out, purged of all traces of contingency, and stuffed with inscriptions so that they can be considered robust and incontrovertible results. Latour and Woolgar call this a 'splitting and inversion' process whereby:

a   an object is separated – and thus acquires a life of its own – from the statements about it: justification for the statement 'AIDS is caused by the HIV virus' no longer needs a basis in experimental evidence or results, but ensues from the *fact* that 'AIDS is indeed caused by the HIV virus' (*splitting*);
b   the research process is reversed: the relation between HIV and AIDS has always existed; it was only waiting to be discovered (*inversion*).

Thus, Knorr-Cetina distinguishes between the 'informal' reasoning which characterizes the laboratory and the 'literary' reasoning that informs the writing of a scientific paper. Far from being a 'faithful' report on the completed research, a paper is a subtle rhetorical exercise which 'forgets much of what has happened in the laboratory' and reconstructs it selectively. For example, a researcher may find him/herself studying a certain problem or using a certain method for reasons which are relatively fortuitous or dictated by the availability of certain resources. But the process will be rationalized in the paper, and the researcher's every move will be made to ensue systematically from specific objectives fixed at the outset.

The two principal sources used by Garfinkel to analyse the discovery of the pulsar by the group of astrophysicists – on the one hand their conversations and the notes jotted down during their observations, on the other the official paper in which the discovery was presented – differed substantially. The work materials revealed a laborious process of successive approximations, adjustments, elaborate discourse practices and common-sense arguments by which the researchers reached agreement on the meaning of what they had observed. But in the article that the astrophysicists published, all this disappeared, being replaced by a presentation of the scientific fact – the pulsar – as 'natural' and independent of any intervention by the observers: a sort of a posteriori rationalization which carefully removed any semblance of 'local historicity' from the process.

> The pulsar is depicted as the cause of everything that is seen and said about it; it is depicted as existing prior to and independently from of any method for detecting it and every way of talking about it.
>
> (Garfinkel *et al.*, 1981: 138)

In entirely similar manner, Gilbert and Mulkay have analysed numerous conversations, discourses and texts by scientists to identify two rhetorical repertoires. The first, what they call the 'contingent' repertoire, dominates informal discussions, laboratory work, notes and intermediate accounts; the second, the empiricist repertoire, is used in every form of official presentation, from a conference paper to the official speech made by the scientists when receiving an award (Gilbert and Mulkay, 1984).

Although the laboratory studies approach does not deny that scientific activity tends to standardize methods and procedures, an aspect constantly stressed is the strongly local and idiosyncratic character of the procedures by which a scientific fact is created. Every experimental setting, every laboratory, even the performance of the same experiment by different researchers, is characterized by a specific pattern of skills, manual techniques and materials.[2] Apparently insignificant events like the escape of a laboratory guinea pig may sometimes significantly alter the entire course of a research project. For his celebrated public experiment on the anthrax vaccine, Pasteur had to use sheep instead of the cows that he had planned because the latter were much dearer to the hearts of the farmers who had volunteered to make their animals available for his experiment (Cadeddu, 1987; Bucchi, 1997; see also Section 3, pp. 70ff.).

This aspect marks a result but also a methodological shortcoming of laboratory studies in regard to the generalizability of observations made in specific settings.

However, the criticism most frequently brought against laboratory studies obviously centres on the concept of 'construction of scientific fact'. The extent to which this criticism is justified depends among other things on which version of the argument is selected, because the degree of 'constructivism' varies from author to author – and, indeed, even among studies made by the same author (Hacking, 1999). It ranges from an extreme version according to which 'facts are consequences rather than causes of scientific descriptions' to more moderate versions which claim that 'what does indeed come into existence when science "discovers" a microbe or a subatomic particle, it is a specific entity distinguished from other entities ... and furnished with a name, a set of descriptors, and a set of techniques in terms of which it can be produced and handled' (Knorr-Cetina, 1995: 161).

> Constructionism did not argue the absence of material reality from scientific activities; it just asked that 'reality' or 'nature' be considered as entities continually rentranscibed from within

scientific and other activities. The focus of interest, for constructionism, is the process of transcription.

(Knorr Cetina, 1995: 149)

At a more specific level, it is certainly possible to question the explanatory capacity of these studies. Beyond their undoubted punctiliousness in describing the routine of scientific work, it is not alway easy to discern their ability to explain how this tangle of micro-interactions and negotiations can be unravelled into a set of shared practices and results. In other words, it is not always clear how consensus, or even communication, is possible in a specific sector of research.

It is possible that this limitation is due to the substantially 'intramural' standpoint taken by these studies (Knorr-Cetina, 1995: 162), by which is meant a view restricted to the laboratory and to scientific actors. It would be important from this perspective, for example, to explore how processes of negotiation and construction are tied to the broader social context. If construction of the scientific fact does occur, it is clear that it does not cease with publication of a scientific paper but continues in numerous further settings and with the participation of multiple actors.

The conversation between doctor and patient about an illness, the production of a technology based on a scientific discovery and its use by consumers, the teaching of a scientific theory in a school classroom, the taking out of an insurance policy based on the estimated probability of a certain event: all these situations are integral parts of this construction process, and contribute to making a scientific fact increasingly solid.

It is not entirely a paradox to say that, in this sense, the laboratory studies approach has been scarcely 'sociological', and that it is driven by a theory centred on science's 'internal' processes rather than on its relationship with society. In rejecting the structural approach to the relationships between science and society, and ultimately the distinction itself between science and society, the ethnographers of scientific knowledge render the social dimension more pervasive but at the same time more difficult to identify. Society penetrates the laboratory, but in the form of an invisible gas. As we shall see, scholars engaged in laboratory studies have responded to these criticisms in various ways.

2 Inside the controversy

Attempts to supersede the strong programme also characterize the strand of studies, centred on the so-called 'Bath school' and scholars like Collins and Pinch, which has culminated in the 'empirical programme of relativism'. Although in many respects akin to laboratory studies, this strand of analysis warrants discussion on its own.

In this case, too, the main concern is with contemporary scientific research and the conduct of detailed case studies (Collins, 1975; Collins and Pinch, 1993). One of the distinctive features of this approach, however, is its focus on scientific controversies as affording significant insights into the processes of scientific activity.[3]

In 1969, the physicist Joseph Weber of the University of Maryland announced that he had discovered large quantities of gravitational radiation from space using a detector of his own invention. Some scientists thought that these gravity waves were predictable on the basis of the general theory of relativity, but no one had yet been able to detect them. Very soon, several laboratories had equipped themselves with apparatus like Weber's. But the difficulty of measurement and the presence of numerous disturbance factors prevented corroboration of Weber's findings by other researchers, who confusingly recorded both positive and negative results. The detector measured vibrations in an aluminium bar, but some of these vibrations were due to electrical, magnetic or seismic phenomena. What was the threshold beyond which the radiation could be assumed to be effectively due to gravity and not to these other factors?

The story of Joseph Weber and his experiments on gravity waves is one of the best known cases studied by Collins and Pinch. On interviewing numerous scientists involved in the controversy and analysing communications among them, Collins and Pinch identified a phenomenon that they called 'the experimenter's regress'. In order to decide whether or not the gravity waves existed, the researchers had first to build a reliable detector. But how could they know if a detector was reliable? They could only be certain that they had a reliable detector if they were sure that the waves existed; in which case a detector that recorded them would be a good one; and one that did not would be an unsatisfactory one. And so on, in a vicious circle.

> Experimental work can be used as a *test* if some way is found of breaking into the circle of the experimenter's regress. In most science the circle is broken because the appropriate range of outcomes is known at the outset. This provides a universally agreed criterion of experimental quality. Where such a clear criterion is not available, the experimenter's regress can only be avoided by finding some other means of defining the quality of an experiment: and the criterion must be independent of the quality of the experiment itself.
>
> (Collins and Pinch, 1993: 98)

What criteria were used by the scientists to settle the gravitational radiation controversy? The answer is social criteria: the reputation of the experimenter and his institution, his nationality, status in his particular research community, the informal opinions of his colleagues. Once it had been established what experiments and researchers could be regarded as reliable, it was a simple matter to determine whether or not the gravitation waves existed: 'defining what counts as a good gravity wave detector, and determining whether gravity waves exist, are the same process. The scientific and the social aspects of this process are inextricable. This is how the experimenter's regress is resolved' (Collins and Pinch, 1993: 101).

Thus emphasized is the implausibility of the 'algorithmic model' of scientific knowledge, whereby experiments and results can be universally repeated on the basis of information provided by papers and scientific reports. Rather, the replication of experiments is an operation which is anything but straightforward and often rests on complex layers of tacit and informal knowledge. In his study of the attempts by research teams to reproduce a model laser already built in the laboratory, Collins shows how difficult it was for them to build the laser solely on the basis of general technical information in scientific articles. They were only able to do so after a long series of meetings, visits by researchers and technicians to other laboratories, and exchanges of material and apparatus (Collins, 1974). The transfer of concepts and methods from one research setting to another – a feature also highlighted by historians of science[4] – is often only possible when researchers change disciplinary areas.[5]

On the basis of similar studies, Collins has drawn up a 'manifesto' which has taken the name of the 'empirical programme of relativism' (Collins, 1983). This programme sets itself three main objectives:

a   to demonstrate the 'interpretive flexibility' of experimental results, i.e. the fact that they may lend themselves to more than one interpretation;
b   to analyse the mechanisms by which closure of this flexibility is achieved – and therefore, for example, the mechanisms by which a controversy is settled;
c   to connect these closure mechanisms with the wider social structure.

As shown by the case of the gravity waves, in the absence of a theoretical framework and a shared technical culture, the mechanisms that enable closure of a controversy and consensus on a certain interpretation may be social in nature: the reputation of a certain scientist, the ability of one particular research group to impose its view of the facts or its own apparatus upon the others: 'It is not the regularity of the world that imposes itself on our senses but the regularity of our institutionalized beliefs that imposes itself on the world' (Collins, 1985: 148).

It should be pointed out that not all actors and scientific institutions have equal importance in this regard. There is, in fact, a 'core set' of researchers and scientific institutions within the broader community which possesses particular resources and a key position in the network for use in orienting the solution of a controversy in a certain sector.

The controversy on gravity waves dragged on for six years amid conflicting results, until a particularly influential researcher joined the fray and was able – albeit by means of a highly questionable experiment – to catalyse criticisms of Weber's original results.

We may now summarize the main features of Collins' and the Bath school's approach. Collins declares that he rejects at least two principles of the strong programme: that of reflexivity, which he believes to be inapplicable, and especially that of causality. Collins is not interested in abstract discussion of the causal relationship between the social dimension and scientific practice; his concern is, instead, (even more forcefully than laboratory studies) to embed the former in the latter, inserting it through the breach opened up by interpretive flexibility.

Nevertheless, he believes it of vital importance to explore and expand the symmetry principle. The sociologist who studies a controversy must be indifferent to its final outcome, to the point that 'the natural world must be treated as it did not affect our perception of it' (Collins, 1983: 88). It is not completely clear to what extent this extreme methodological relativism translates into a substantial relativism also at the epistemological level, as it may appear from some of Collins' writings (1981, 1985).

Although it is to some extent justified, the decision to study controversies as a particularly rich source of data for the social analyst is a methodological choice of no little account. It also seems that Collins subscribes at least in part to an 'agonistic' and rationalist model of scientific debate, where two sides battle it out until one prevails over the other. But science, and contemporary science in particular, offers numerous examples of research sectors which are much more fragmented than this and in which different positions overlap. In certain cases – for instance the botanists studied by Dean (cf. Chapter 2) – the positions may be so distant and irreconcilable as to prevent communication itself, and therefore preclude any settlement of the controversy. One also thinks of the debate provoked by the famous article in which Alvarez and his colleagues attributed the mass extinction of the dinosaurs to extraterrestrial causes, like the Earth's collision with asteroids. The generality of Alvarez's hypothesis, its importance for various scientific sectors (statistics, geology, palaeontology and astrophysics) and its resonance in terms of images and metaphors (the combination of two mysteries, one in space one on earth; the extinction of the dinosaurs as a metaphor for human extinction, because the statistical models used to explain the extinction were taken from research on nuclear weapons; the word 'extraterrestrial' used in the title of the original article, which evoked Martians more than asteroids) have stoked the debate for the past 15 years (Clemens, 1986, 1994). Scientists have different *opinions*[6] on the matter according to their particular perspective – just as they would on other topics of public interest – and from time to time evidence is produced either to confirm or confute the hypothesis.

Finally, this approach, too, seems largely to neglect the role that actors external to the 'core set' and the scientific community can play in settling a controversy, and more generally in scientific debate. In the already-mentioned case of Pasteur's anthrax vaccine, the support of veterinarians, farmers and journalists was crucial for Pasteur's ability to overcome the extreme reluctance of his colleagues to accept that a disease could be prevented by inoculation with the same infective agent (Latour, 1984; Bucchi, 1997). More recently, the role of non-experts – activists or representatives of patients' associations – in the definition of research protocols in regard to medicine and the environment has become so massive and pervasive as to be institutionalized into panels where lay citizens sit together with scientists and policy makers.[7]

3 Science as a two-faced Janus: actor-network theory

Actor-network theory can be viewed as an attempt to expand the explanatory capacity of the microsociological approaches to science discussed thus far. Developed by a group of scholars headed by Bruno Latour and Michel Callon, actor-network theory takes up the argument where laboratory studies left off.

For the proponents of this approach, science has two faces, like the Janus of Roman mythology: on the one hand there is 'ready-made' science; on the other, science 'in the making' or research. While it is the task of epistemology to analyse the characteristics of the former, it is the task of the sociology of science to study the latter.

Inside the laboratory 71

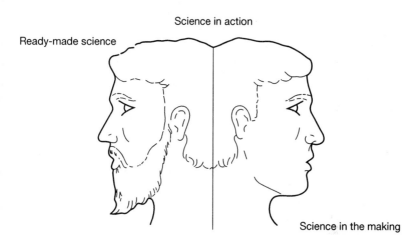

Figure 4.1 Science as a two-faced Janus
Source: Latour (1987: 4)

By entering this 'side door', the sociologist can examine the processes that lead to construction of a scientific fact. Solid form cannot be given to a scientific fact without the support and cooperation of an entire series of 'allies' both within and without the laboratory. A scientific statement or a finding can only acquire the status of 'fact', or conversely of 'artefact', if a complex network of actors – beginning with research colleagues who cite your findings or criticize them – pass it from hand to hand.

> A statement is thus always in jeopardy, much like a ball in a game of rugby. If no player takes it up, it simply sits on the grass. To have it move again you need an action, for someone to seize and throw it ... the construction of facts, like a game of rugby, is a collective process.
> (Latour, 1987: 104)

'The fate of what we say and make is in later users' hands', argues Latour (1987: 29), concluding that 'the construction of facts and machines is a collective process'. In order to depict this support network, Latour begins by disputing a series of distinctions.

The first of these distinctions is that between science and technology, which Latour replaces with the synthetic term 'technoscience'. The feature shared by a scientific finding or a technological object is that they are both 'black boxes'. The term, borrowed from cybernetics, denotes a mechanism which is too complex for its analysis to be possible, with the consequence that one is forced to settle for knowledge of only its input and output. This is what happens to a scientific finding or a technological object once they have become established: they are cited and utilized without being questioned further or 'dismantled'.

The second disputed distinction is that between human and non-human actors. A research colleague, a bibliographical citation in a paper, an apparatus which yields a microscope image, a company willing to invest in a research project, a virus that behaves in a certain way, the potential users of a technological innovation: all these are allies in the process that transforms a set of experimental results and statements or a technological prototype into a 'black box': a scientific fact or a technological product.

Latour cites the example of the *Post-It* note. Initially considered a failure by the company that produced it, 3M, because it wanted to produce a strong glue, the easily detachable *Post-It* note was, instead, proposed by its inventor as a useful device to mark book pages without dirtying them. In order to overcome the resistance of the marketing department, a quantity of *Post-It* notes were distributed to the company's secretaries, who were told to ask marketing for more when they had finished their supplies.

The most celebrated case examined by Latour is that of Pasteur and his discovery of preventive vaccines. Latour says that he wants to represent this discovery as a sort of *War and Peace*, showing that Pasteur's victory was not solely the result of his genius but was also brought about by a complex network of alliances and troops supporting 'General' Pasteur. Opposed by many of his colleagues because of his explanation of infectious diseases and his hypothesis – deemed absurd – that they could be prevented by inoculations with the same disease, Pasteur was able to construct his scientific fact by enlisting the support of veterinarians, hygienists, farmers, and even of the bacteria themselves!

However, this process is far from being straightforward for whenever a new supporter enters the network, the scientific statement or the technological artefact is stretched and adapted to accommodate different interests. The key concept is that of 'translation' or 'the interpretation given by the fact-builders of their interests and that of the people they enrol' (Latour, 1987: 108). As in a process of military enrolment, potential allies must be persuaded that supporting the scientific fact is in their interest.

For example, Pasteur was able to translate some of the problems besetting the French farmers of his time into bacteriological terms, and thus present his work as being in their crucial interest: 'If you wish to solve the anthrax problem, you have first to pass through my laboratory', he wrote. His laboratory thus became an 'obligatory point of passage', and Pasteur was no longer alone in fighting his battle. Translation enabled him both to enrol allies and to retain control over his own 'fact'.

How, asks Latour, can one explain the fact that after 20 years of hostility towards Pasteur's discoveries and methods, doctors suddenly 'became enthusiastic about his research, asked for courses in bacteriology, transformed their surgeries into small laboratories, and became experts and ardent propagators of anti-diphtheria serums' (Latour, 1995: 29).

The crucial 'translation' in this case was the transformation of vaccines into serums, especially following the discoveries by one of Pasteur's pupils, Roux. Whereas the doctors had complained that preventive vaccines 'deprived them of work', because they reduced the number of patients and introduced competition by hygienists and vaccinators, serums could be easily incorporated into medical practice because they required diagnosis and the a posteriori administration of a substance entirely similar to any other drug. In this manner, the doctors and the Pasteur Institute became reconciled, to their mutual advantage.

Latour's description of this complex process is his reply to the question that we have seen traversing science studies since Kuhn: how is it possible to explain the transition from one paradigm to another and, more generally, how is it possible to explain the evolution of scientific ideas? Latour invites us to abandon the traditional 'diffusion' models whereby a scientific finding or a technological

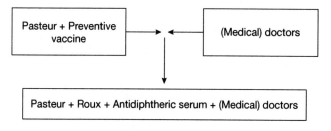

Figure 4.2 Translation and success of Pasteur's vaccine
Source: Latour (1995: 31)

innovation is able to propagate itself under its own impetus, with no need for any other assistance. This model, Latour claims, can only survive if we emphasize 'exceptional' factors like the presence of pioneers or great scientists working in isolation. Yet, the diffusion model is not even able to provide a satisfactory explanation for the change of attitude towards a discovery or an innovation: consider, for example, the initial resistance raised by doctors against Pasteur's discoveries and against vaccination in general.

From these considerations, Latour derives two methodological rules that challenge not only a naturalistic view of scientific research but also a large part of the social studies of science conducted hitherto.

'Since the settlement of a controversy is the cause of Nature's representation not the consequence, we can never use the outcome – Nature – to explain how and why a controversy has been settled' (Latour, 1987: 99). In other words, if 'black boxes' – scientific facts – result from the complex mobilization of diverse supporters, we cannot use black boxes for the purpose of explanation. Roux's antidiphtheria serum or Pasteur's vaccines are not the initial datum but the result of a process. It is therefore wrong to say that it was the serum which convinced the sceptical doctors, or that it was the *Post-it* which convinced the marketing managers at 3M that it was marketable.

Latour admits that, although his principle is applicable to current, present-day controversies, it is much less able to account – in a historical perspective – for those already closed. While it is easy today to use present knowledge in physics to argue that the Blondlot fiasco was caused by the non-existence of N-rays, it was not so easy to contend thus at the time of the controversy. However, when the black box is shut – so that the network of alliances that have supported it disappear from view – the cost and difficulty of re-opening it are excessive for any actor, including the historians and sociologists of science. At this point 'Nature talks straight, facts are facts'. This methodological rule, with its combination of realism and relativism, strikes Latour as a 'good balance' which enables us 'to trace with accuracy the sudden shifts from one face of Janus to the other. This method offers us, so to speak, a stereophonic rendering of fact-making instead of its monophonic predecessors!' (ibid.: 100).

Latour's second methodological principle may be of more interest to those engaged in analysis of knowledge from the sociological point of view. Society as a specific dimension with respect to science and technology plays a key role in the traditional diffusion model: when the diffusion or acceptance of a fact or an object ceases, social factors may be invoked. For Latour, indeed, the very 'belief in the existence of a society separated from technoscience is an outcome of the diffusion model' (ibid.: 141).

If we cannot use Nature as the reason for solution of a controversy, neither can we use Society, because the stabilization of alliances and of social interests is the result of the controversy, not its starting point.

This is not the place for detailed discussion of the wide-ranging debate aroused by actor-network theory. Chapter 7 will resume some of the themes touched upon here.

The criticisms brought against actor-network theory are of essentially two kinds. The first is more general and 'external' and concerns the explanatory capacity of the approach, which is accused of being tautological. If interests and allies are translated and enrolled but not persuaded by scientific-technological contents, it is unclear which mechanisms lead to success and which instead lead to failure. What was it that enabled Pasteur to win? The network that Latour plots around Pasteur's discovery seems to fragment the role of various factors, rather than constructing an alternative explanation. His admission of the difficulty of studying already closed controversies seems to justify this criticism to a certain extent (see Amsterdamska, 1990).

The second, and more specific, criticism concerns the idea that certain scientific actors are able to control the entire process by means of a 'Machiavellian' and preordained strategy. Studies conducted on the perception of science and technology have shown that other actors may appropriate a fact and radically adapt it to their purposes. Moreover, an ally's membership of a certain network is often erratic and subject to a complex set of circumstances such that it becomes difficult – even for a well-positioned actor like Pasteur – to maintain complete control over the situation.

The Big Bang theory of the origin of the universe has already been mentioned. Considered to be only one of the various explanations available for the birth of the cosmos and, until the mid-1960s not even the most accredited of them, the theory owes its name to one of its fiercest opponents, the astrophysicist Fred Hoyle. One of the best-known scientists of the post-war period, Hoyle delivered a celebrated series of popular science lectures on BBC radio. During one of these broadcasts, on finding that he had to refer to the theory rivalling his own (Hoyle was one of the original theorists of the 'steady state theory', which held that the universe has never been in a state of singularity, i.e. has never had an origin), Hoyle disparagingly dismissed it as the 'Big Bang theory'. So graphic was his epithet, however, that it became the standard label for the theory even among specialists; and it persuaded public opinion that the Big Bang was the origin of the universe well before experimental evidence provided important confirmation of the fact. Thus, Hoyle's insult rebounded against himself and his theory (Gregory and Miller, 1998).

Notes

1 For a general overview of ethnomethodology studies see Heritage (1984).
2 This aspect is often denoted by the term 'indexicality' introduced by Pierce and used by ethnomethodologists. 'Indexicality' refers 'to the fact that the meaning of every account, verbal or otherwise, is tied to the setting in which it is produced' (Giglioli and Dal Lago, 1983: 17).
3 Albeit in different form, the usefulness of studying disputes in the scientific community – for example over who has been the first to make a discovery – was first pointed out by Merton (1973).
4 See for example Fox Keller (1995).
5 See Mulkay (1974).
6 This is the term used by the palaeontologist David Raup in his reconstruction of the debate (Raup, 1991).
7 Emblematic in this regard is the case of AIDS research (see Epstein, 1996). See also Callon and Rabcharisoa (1999).

5 Tearing bicycles and missiles apart
The sociology of technology

1 The importance of a stirrup

According to a celebrated historical study, the advent of feudalism was due to the invention of the stirrup. The use of stirrups radically reduced the likelihood that armoured knights would fall off their horses during combat; it increased their efficiency in battle and thus, the theory ran, fostered the rise of feudal society based on landownership and the use of force (White, 1962).

This type of analysis is a good example of the approach known as 'technological determinism', which takes the development of technology for granted and then merely examines its impact on the economy and society. Under this approach, the sociology of technology largely restricts itself to analysis of the social consequences of technological development. In the case just cited, the introduction of a specific innovation, namely the stirrup, is viewed as the cause of so profound a historical change as feudalism. Without denying that this aspect is of considerable importance, the majority of studies conducted by sociologists of technology in the past 30 years have marked out other and more meaningful areas for the sociological analysis of technology.[1]

Far from representing a simple appendix to the sociology of science, this line of inquiry has developed and ramified to such an extent that the whole field is now customarily summed up by the abbreviation STS – Science and Technology Studies. Although somewhat overshadowed by the more strident debates on the sociology of scientific knowledge, the analysis of technology has been part of the discipline's history since its beginnings. One thinks of Merton's pioneering study on 'Science, Technology and Society in Seventeenth-Century England', which disputed the deterministic relationship between economic development and the institutionalization of scientific practice, emphasizing the importance of socio-cultural factors like Protestant values (Merton, 1938). Some of the most innovative approaches and most significant studies of recent years in the sociology of science make much use of empirical data and case studies from the technological sphere, to the point that they call into question – an example being Latour and actor-network theory – the distinction itself between science and technology, replacing it with the more general concept of 'technoscience'.

After all, compared to science, technology is something with which we have much more frequent contact in our daily lives: from telecommunications to ultrasound scans, we constantly encounter technology. The theme of technology enables the analyst to highlight the intersection among different disciplines – those of historians, epistemologists, sociologists, economists and anthropologists – characteristic of the most fruitful periods of study on science and technology.

2 The clockmaker who astonished the astronomers

In the summer of 1730, the clockmaker John Harrison presented himself at the Royal Observatory of Greenwich carrying a mysterious wooden box. He asked to speak to Edmund Halley, the Observatory's director. Halley, as Astronomer Royal and already celebrated for his observations on the moon and the motion of the stars, was not only the head of the Observatory but also a leading member of the Board of Longitude. Set up sixteen years previously by Queen Anne, this Board administered a conspicuous prize of 20,000 pounds (some millions of today's pounds) to be awarded to whoever solved the problem of calculating longitude. The longitude problem was of extreme importance at the time. The difficulty of accurately fixing a ship's position was the cause of innumerable accidents and shipwrecks, and finding a solution to the problem was crucial, for both military and commercial reasons. The greatest astronomers of the age had wrestled with the problem, propounding complex solutions based on the eclipse of Jupiter's satellites, or the diurnal distances between the moon and the sun and the nocturnal ones between the moon and stars. In his coarse English of the countryside, Harrison showed an intrigued Halley the contents of his box. It was a timepiece which would function – thanks to special devices – on board any vessel and in all atmospheric conditions. Harrison had solved the longitude problem, and it was now possible for seafarers simultaneously to know the time at their point of departure and the local time of their ship. Harrison spent the rest of his lifetime perfecting his invention and battling against the astronomers of the Board, who, as Halley had predicted, were loath to accept 'a mechanical answer to what (they) saw as an astronomical question' (Sobel, 1995: 75). From the end of the 1700s onwards, however, there was no ship's captain anywhere who did not possess a chronometer of the type invented and built by Harrison.

The story of Harrison and his solution of the longitude problem controverts the assumption associated with technological determinism to the effect that technology is solely an applied science. On this assumption, only science drives technological innovation, which is nothing but the automatic application of scientific discoveries. This image of science as the 'goose that lays the golden egg' has played a decisive historical role in the recognition of the importance of public support for basic research and the autonomy of the scientific community, especially since the Second World War. One of the first government policy documents on research policy, the report prepared by Vannevar Bush for American President Roosevelt, and significantly entitled *Science: The Endless Frontier* (1945), propounded a similar view of science. According to the report, scientific research had proved itself amply able to furnish economic and practical benefits for society as a whole. 'Basic research leads to new knowledge', Bush wrote in his report,

> it creates the fund from which the practical applications of knowledge must be drawn. New products and new processes do not appear full grown. They are founded on new principles and new conceptions, which in turn are painstakingly developed by research in the purest realms of science.
> (Layton, 1977: 206)

It was for these reasons that the Bush Report called for generous and long-term support for research, while respecting both the autonomy of scientists and their ability to determine on their own the areas of research most deserving investment of money and human resources (by means of the peer review process: that is, the assessment of a researcher's work by his/her fellow researchers which is now standard practice in scientific inquiry). This ideal was institutionally embodied in the National Science Foundation (NSF), created in 1950 on the proposal of Bush himself. A more markedly practical document, *Science and Public Policy*, better known as the *Steelman Report* (1947), urged that expenditure on research should be doubled over the following ten years.

However, at the end of the 1960s, the central role of scientific research in technology and economic development began to come under dispute. A wide-ranging report called *Project Hindsight*, commissioned by the Department of Defense, engaged technicians and engineers for fully eight years in the analysis of 20 apparatuses vital for the security of the nation. Identified for each of these apparatuses were a series of 'events' that had made their development possible. These events were classified as either 'technological' or 'scientific' and then further distinguished between 'applied research' and 'basic research'. It was found that 91 per cent of the significant events were technological, 9 per cent were scientific, and only 0.3 per cent could be characterized as basic research. The scientific community saw the danger and responded with another study called TRACES, commissioned by the National Science Foundation. This study, which used methods entirely similar to those employed by the previous one, examined ten technological innovations of particular significance, but it obtained entirely different results: 34 per cent of the events considered in relation to these innovations came from basic research, and 38 per cent from applied research.

How can one explain this marked difference between the results of studies conducted in the same years, with similar methods, and often on the same technological innovations? It would be too easy to attribute the contradiction between the two reports solely to the differing institutional purposes of their commissioners. But it is likely that one cause was the difficulty of classifying an event as either scientific or technological.

One of the distinctive features of contemporary science, in fact, is its increasing overlap with technological development, so that scientists work in typically applied sectors while engineers engage in research. Since the early 1950s, it has been commonplace for the American universities in Silicon Valley to recruit their lecturers in solid state physics from staff working in local electronics companies (Rosenberg, 1982). The possibility for scientists to conduct their own research – especially in sectors like particle physics – depends increasingly on the contribution of technicians: for example, 30 per cent of the personnel at CERN – the world's largest Particle Physics Laboratory – consists of researchers and 60 per cent of technical staff.

It is no longer science that stimulates technology in this interaction. Technology also influences science, identifying sectors or topics for fruitful scientific research, or furnishing the apparatus that makes certain experiments and observations possible. Historians of science, moreover, have shown that the relationship with technique and the manual arts was one of the factors crucially responsible for the birth of modern science (Rossi, 1988b). Not only have some of the most important innovations in the history of technology, like the spinning jenny or the steam engine, resulted from the application of scientific discoveries, but in some cases it has been technological innovations that have had a significant impact on science. The problems encountered and the solutions adopted by technicians to develop motors prompted the reflections which led Carnot to formulate his general principles of thermodynamics (Barnes and Shapin, 1979; Layton, 1988).

One of the most striking examples of scientific research stimulated by technical activity is the discovery that won the Nobel prize for Arno Penzias and Robert Wilson. Penzias and Wilson were working as technicians at Bell Laboratories, their task being to solve the problem of disturbance on telephone lines. Using a highly sensitive antenna to capture signals from one of the first telecommunications satellites, they were unable to eliminate a background noise which persisted whatever device they used, and whatever direction they pointed the antenna. In the end, they came to the realization that the noise could only be the cosmic background radiation predicted by Gamow's theory on the origin of the universe: the echo of the primordial Big Bang.

No less simplistic is the assumption that technological innovation results from a pure act of individual creativity by a single inventor – the 'heroic' figure personified by geniuses like Franklin or Edison. This assumption must take account of aspects similar to those already discussed as regards science. First, an innovation frequently arises within a particular technological paradigm, or in other words, within an already-existing framework of resources, models and technologies: for instance, the missiles developed by the US and the Soviet Union after the Second World War were modelled on the German V-2 rocket (Mackenzie and Wajcman, 1999). Moreover, technologies tend not to arise in isolation from each other but are instead embedded in broader technological systems. The technology of the television set, for example, presupposes a technology for the transmission of images by means of radio waves, relay stations and antennas. A technological innovation is also part of broader economic and social systems. In his classic study on Edison's invention of the electric light bulb, Hughes shows that what induced Edison to concentrate on a filament with sufficient resistance, so that he could increase the voltage while reducing the current, was a set of economic-organizational constraints: the need to keep the cost of electricity low, to deliver it to a large number of consumers in like manner to gas, and to take account of the high cost of copper (Hughes, 1999). The technology of the Italian high speed trains – the *Pendolini* – which differs radically from that employed in other countries, was developed in a technological, economic and political system characterized by obsolete infrastructures and by the rigidity and slowness of the procedures to innovate it; this set of constraints shifted the focus of the innovation to the vehicle itself.

To return to my initial example, the adoption of the stirrup could only produce its particular effects within a certain economic, political and cultural setting. It did not have any significant impact on Anglo-Saxon society, for example, until after the Norman Conquest (Mackenzie and Wajcman, 1999). Finally, a technological innovation does not always result from intentional and linear processes; rather, it often arises from the coincidence of a variety of forces and social actors. The personal computer, for example, was not born in the IBM laboratories – 'I think there is a world market for maybe five computers' said the company's chief executive officer, Thomas J. Watson, in 1943 – but in the basements where youngsters tinkered with electronic equipment so that they could make free telephone calls (Ceruzzi, 1999).

3 A mysterious cyclist

In June 1881, while sojourning on the Isle of Wight, Queen Victoria saw from her carriage a young woman travelling at considerable speed on a curious contraption. The Queen ordered one of her attendants to track down the girl, who shortly afterwards presented herself at the royal residence astride a tricycle sold by her father, the only dealer on the island. What was so special about this tricycle that it should arouse the interest of the Queen of England? And how in the space of a few short years did it turn into the bicycle that we know today? In other words, how does a technological device develop and spread?

We already know the answer offered by technological determinism: it is the 'best' device that imposes itself by virtue of its efficiency. However, the technological device which is the most efficient from its user's point of view may not be so from others. For example, a certain piece of equipment may suit the needs of the employer but not those of his/her employees, or it may not meet the standards required by environmental protection. The capabilities of the cell phone may have been adequate for the first generation of its users, but when the device became an object of mass consumption, it had to be simplified. The emphasis was now on certain functions – like text messaging – which were of entirely marginal importance to the cell phone's first users. The case of musical synthesizers is exemplary. The first electronic musical synthesizers were complex and costly pieces of equipment produced for musicians with classical training who were interested in experimentation. In his spare time, an engineer at the Moog company assembled some simple modules with a keyboard. But who would have been interested in a synthesizer of this kind, musically more limited than the ordinary electronic organ? Not the usual purchasers of Moog instruments, but certainly the new category of users consisting of rock musicians, who were looking for an easy-to-use instrument whose sounds, however limited, could be modulated even during a live performance (Pinch, 1999).

A further element should be borne in mind. The various phases of the innovation process often form an uninterrupted sequence, so that it is difficult to separate the innovation stage from the diffusion stage. Economists of innovation talk of 'learning by using' with reference to improvements made to a device, not during its production but during its use (Rosenberg, 1982). These improvements may or may not be 'incorporated': in other words, they may either give rise to a physical modification of the innovation, or simply to a change in its use. The experience of users of turbojet aircraft prompted development of maintenance procedures and flying techniques which encouraged their purchase, improved their performance, and induced engineers to redesign certain components. In the same way, the difficulties encountered by numerous users of video recorders have prompted the manufacturers to make modifications to them.

It is, accordingly, clear that earlier entry onto the market by one technology rather than another often gives the former a decisive competitive advantage: however perfectible a technology may be, its dominance increases with the number of people who use it. For example, the distribution of the keys on a PC keyboard (called 'qwerty' after the first six letters in the top row) is no more efficient than any other arrangement, and it is not due to any particular technical exigency. It derives, in fact, from the technology of the typewriters that PCs replaced. The letter distribution on the old-style typewriter served to reduce the jamming of the hammers when adjacent keys were struck in too rapid a sequence. The same goes for operating systems or word processing software (Mackenzie and Wajcman, 1999). The greater length of the tapes used allowed the VHS video recording system to outpace the Betamax – equivalent to the VHS system from a strictly technical point of view – in a sector where the rapid adoption of a universal standard was crucial

(Liebowitz and Margolis, 1995). Innovations like the digital cassette or the laser disc in hi-fi technology had indubitable advantages in terms of sound quality. But they encountered consumers who had absolutely no intention of making yet further investments in terms of both money and 'learning by using', when they had only just spent considerable sums on buying compact disc players. Conversely, a low-fidelity music player like MP3 – a compression format which shrinks audio files by eliminating sounds irrelevant to the human ear – acts on a crucial element of the technological system by significantly reducing on-line download times and therefore telephone bills.

But let us return to the history of the bicycle, a case analysed by scholars working within the framework of the 'social construction of technology' approach (Bijker *et al.*, 1987; Pinch and Bijker, 1990; Bijker, 1995). Abbreviated to SCOT, this approach is articulated into three phases, in close analogy with the 'empirical programme of empiricism' examined in the previous chapter:

a demonstrating the 'interpretative flexibility' of technological devices: the same artefact may be designed in different modes and forms, there is no single optimal solution;
b analysing the mechanisms by which this interpretative flexibility is 'closed' at a certain point and an artefact assumes a stable form;
c connecting these closure mechanisms with the wider sociopolitical milieu.

The overall aim of this approach is to go beyond reconstruction of technological innovation by 'hindsight', so that every artefact results from a necessary sequence of attempts which logically yields the most efficient model, and where all that matters are the technical properties of artefacts. On this view, the history of the bicycle is nothing but 'a simple genealogy extending from Boneshaker to velocipede to high-wheeled ordinary to Lawson's bicyclette, the last labelled "the first modern bicycle"' (Bijker, 1995: 50). In his study of the bicycle, Bijker also examines models that were apparent 'failures', representing the entire course as a multilinear process involving not only bicycle designers and manufacturers but also social groups of users, like cycling clubs and women.

An artefact like a bicycle, therefore, also results from negotiation among social groups. It must resolve problems that these groups regard as being in need of solution; its characteristics are not given

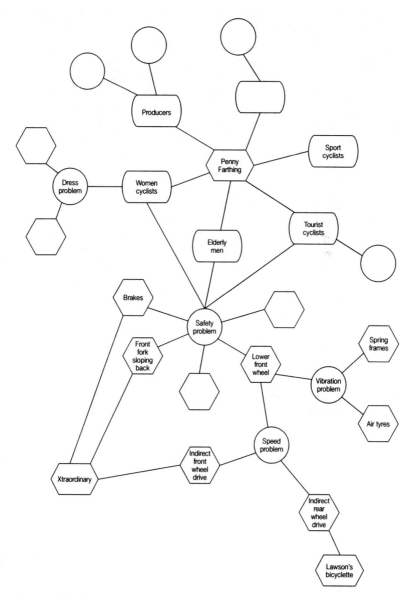

Figure 5.1 Relevant social groups, problems and solutions in the evolution of the modern bicycle

Source: Bijker (1995: 53)

once and for all by the manufacturer but are subject to extreme 'interpretative flexibility' by the actors involved. In Victorian England, there were at least three devices that could legitimately aspire to becoming 'the' bicycle: the high-wheeled ordinary bicycle, the low-wheeled safety bicycle and the tricycle. The high-wheeler or ordinary bicycle was preferred by sportsmen because it gave them a chance to show themselves off as athletic and adventurous. They dismissed the low-wheeled bicycle as a machine for 'sissies'. However, when the low-wheeled bicycle was redefined as a means of transport, as opposed to a device with which to flaunt macho prowess – also because of the greater use now being made of the machine by other social categories (women, for example) – the ordinary bicycle was perceived as more dangerous than the safety bicycle.

Tricycles and bicycles with side saddles meant that women could pedal while wearing long skirts. As a consequence, these models enjoyed a certain amount of success. However, it was realized that the modern bicycle could also be ridden by women when an alternative solution to the modesty problem was found: bloomers worn under a short skirt.

The new model also found favour with sports cyclists because of the invention of another technological artefact: the tyre. Initially a solution for the problem of vibration (and therefore of little attraction to sports cyclists, whose chief source of enjoyment was the thrill of the ride and who cared nothing about vibration), the tyre was then successfully redefined as a means to solve the problem of the bicycle's slowness. However, 'the technologies needed to turn the 1860 low-wheelers into 1880 low wheelers, such as chain and gear drives, were already available in the 1860s' (Bijker, 1995: 97). In the meantime, complex interpretative negotiation had taken place on definition of the main problems and the acceptable solutions, until what Bijker calls 'stabilization' and 'interpretative closure' came about.

As in the scientific controversies studied by Collins and Pinch (see Chapter 4), there comes a point when one of the many interpretations available prevails: the high-wheeled ordinary bicycle is dangerous, full stop. In this sense, the artifacts 'ordinary bicycle' or 'high-speed tyre' are social constructs, in that they result from a process of closure and stabilization which imposes one of the various possible perceptions of the same device (dangerous or 'macho' in the case of the ordinary bicycle; efficient or 'sissy' in the case of the low-wheeled one) held by the social groups involved. Analysis of technological devices must therefore apply the same principle of symmetry as developed by SSK for the study of scientific controversies, adopting an impartial perspective on the efficacy or inefficacy of a machine. This perspective is not given from the outset but results from negotiation among the social groups involved, and from the subsequent stabilization and interpretative closure. Hence, technological 'failures' are just as sociologically interesting as 'successes': a futuristic model of an 'intelligent' underground railway with a system of modular carriages, so that passengers would not have to change trains to reach their destinations, failed to incorporate the conflicting requirements of technicians, managers of the manufacturing company and the Paris city council (Latour, 1992).

One limitation of this approach is the difficulty of identifying all the groups of actors involved in the construction of a particular artefact. Moreover, while the SCOT approach has the merit of emphasizing the role of users in the innovation process, it tends to attribute to all the groups involved the same capacity to influence the closure of the interpretative possibilities. This aspect is indubitably due to the approach's strict descendancy from the sociology of science – and from the empirical programme of relativism in particular (see Chapter 4) – with which it shares an interest in controversies and concepts like interpretative closure.

But while the study of scientific controversies deals with a relatively homogeneous group (researchers engaged in the study of a particular phenomenon), this is not always so in the technological domain. Indeed, it is likely that sports cyclists, cycle tourists, Victorian ladies and gentlemen formed groups of different sizes and organization. It is especially difficult to argue that users on the one hand, and designers/manufacturers on the other can contribute in the same way to the closure process. The interpretative possibilities of the artefact's users, in fact, are largely restricted by the technological characteristics of the device as it appears on the market. As in the case of the empirical programme of relativism, the emphasis on controversies and on the closure process seemingly leads to overgeneralization.

Using another case of a cycling artefact, the mountain bike, Rosen has shown that the distinctive feature of this kind of bicycle is the constantly changing design of its frame. In this case, too, the connection between the micro level of the specific controversy and the wider social context is not explained satisfactorily. The characteristics of the various groups, their differences in terms of prestige and power, their motives, and their places in the social and cultural scenario are not spelled out but are, instead, taken as given. In other words, however ironic it may seem, SCOT 'doesn't explain the social aspects of technological development as richly as the technological aspects' (Rosen, 1993: 508).

According to Rosen, the third stage of the SCOT approach ('connecting the closure mechanisms with their socio-cultural context') can be more usefully conceived as cutting across the first two stages, because it enables definition of the influential social groups, the relevant artefacts and possible closure mechanisms. In his study of the mountain bike, for example, Rosen hypothesizes that the constant variations in its design have been due to changes in cycling culture and, more generally, 'in the post-Fordist economic system to which the cycle industry belongs' (Rosen, 1993: 493).

4 Beyond innovation: what really happened in the skies above Baghdad?

If SCOT approaches analyse the role of the social dimension in innovation, it is evident that innovation does not exhaust all the aspects of technology. What can sociology tell us about these further aspects? In the past ten years, a number of the scholars already mentioned in this and previous chapters have sought to apply the tools of the sociology of scientific knowledge to technological devices. Their intention has been to re-assess the analysis of technology, which has too often been regarded as little more than an appendix reserved for the applied dimension of science. In reality – argue Collins and Pinch – technology enables one to focus on 'the problems of science in another form' – a form perhaps more concrete and better suited to sharpening the focus on the social dimension (Collins and Pinch, 1998: 2). Technological devices incorporate, and also help to reinforce, social phenomena like racial prejudices: for instance, the technologies specific to photography, cinematography and television were developed to reproduce white skin tones, making the filming of black people 'problematic' (Dyer, 1999).

What can the sociology of scientific knowledge tell us about technology that engineers, economists and users cannot? Mackenzie (1996) starts with the problem of how we come to know the properties of technological devices. How do we learn how to make a blender work; or how do we learn how a Patriot missile functions?

Essentially in three ways:

a   by authority: we believe what we are told about these devices by people whom we trust;
b   by induction: we learn the properties of a device by using it and testing it;

c by deduction: we infer the properties of a device from theories and models – for example, from Newtonian physics or Maxwell's laws of electromagnetism.

These three sources of knowledge, according to Mackenzie, are suffused with social elements. In the first case, that of authority, this is quite obvious. Indeed, as trust diminishes, so does cognitive authority. The leaders of the anti-vivisection movement in Victorian England had such little trust in doctors that they rejected traditional medicine entirely, preferring alternative practices like homeopathy. Today, in the same way, if a study proving the safety of genetically modified food were to be published by a multinational with interests in biotechnology, the results would most likely be rejected a priori as unreliable.

As to induction, the similarity relations on which it is based contain an element of social convention (see Chapter 2 and Barnes 1982a, 1982b). Here, I shall concentrate on one particular type of similarity; that between the testing of a technology and its actual use. From this point of view, it is of crucial importance to determine whether and to what extent the test can accurately predict how the device will behave when used at peak regime. In the case of nuclear missiles, the test may consist of launching them without warheads at a Pacific atoll. But to what extent does this exercise show what would actually happen if nuclear missiles were launched from – say – Dakota and aimed at Moscow?

In the US, especially during the 1980s, fierce controversy erupted among experts when it was claimed that missile test firings yielded little or no information about the actual performance of nuclear-armed missiles, which in war would have different trajectories and be fired under different conditions. The experts split over the representativeness of the test firings, and their similarity to the real war situation; moreover, positions taken in the controversy revealed a 'clear social patterning' (MacKenzie, 1996: 254) insofar as criticism of inferences drawn from missile testing to use were much more widespread among the proponents of the nuclear bomber aircraft.

But even the testing of a technological device in actual war conditions may prove problematic. This is the case of the celebrated Patriot missiles hailed by the American military command as its key resource in the Gulf War, a weapon able to intercept – according to President Bush himself – '41 out of 42 Iraqi Scud missiles'. Theodore Postol, professor of science, technology and national security policy at MIT, on examining television video footage noted the extreme inaccuracy of the Patriot missiles, which in most cases missed the Scuds, or hit their tanks rather than the warheads. The fact that the Scuds exploded or fell at a distance from their targets was not evidence that they had been intercepted, because the Scud is a missile which is intrinsically unreliable. Although the supporters of the Patriot acknowledged the validity of many of Postol's criticisms, they countered by saying that the frame frequency of his video footage was not rapid enough to handle the extreme speed of the missiles. Fully three different interpretations were given to the video evidence: that of the media, for which explosions in the sky were sufficient proof of interception; that of Postol and other critics; and that of the Patriot's defenders. Congressional hearings and government inquiries conducted on the basis of Postol's documentation produced very different estimates of the Patriot's effectiveness: 'between 42% and 45%'; '90% in Saudi Arabia and 50% in Israel'; '60% overall'; '25% with confidence'; '9% with complete certainty'; 'one missile destroyed in Saudi Arabia and maybe one in Israel' (Collins and Pinch, 1998: 9). Though disputing Postol's claims, the defence experts admitted that, according to the army's own criteria, there was only absolute certainty that only one warhead of one Scud had been destroyed by a Patriot – although this obviously did not rule out the possibility that there had been more successes. The controversy continued for a long time, with the involvement of numerous actors – experts, army officials, representatives from Raytheon (the missile's manufacturer) – whose interpretations stem from different interests: the 'success' of the Patriots during the Gulf War persuaded the armies of countries like Saudi Arabia to purchase them, and it strengthened the hand of those calling for further investments in the development of intercontinental missiles. Their 'failure' provides a useful argument for the critics of 'Star Wars' and the Bush senr administration. Agreement has been difficult to reach because different criteria can be used to gauge the success of the Patriot missile. One of the main difficulties has been establishing what counts as a 'success'. In the course of the controversy, in fact, a series of direct criteria for the Patriot's effectiveness have emerged – from 'all the Scud warheads dudded' to 'some Scuds intercepted' to 'Israeli lives saved' – as well as indirect ones (from 'Israel was kept out of the war' to 'Saddam sued for peace' or 'Patriot sales increased'). During a congressional hearing of 1992, General Drolet maintained that by saying '41 out of 42 Scuds were intercepted', Bush senr did not actually mean that all the Scuds had been destroyed, only that 'a Patriot and a Scud crossed paths, their paths in the sky. It was engaged' (Collins and Pinch, 1998: 19).

These examples have prompted MacKenzie to formulate a more general interpretation of people's attitudes towards technological artefacts, in particular their tendency to cast doubt on them, as did the Patriot's critics, or instead accept them as 'black boxes' ready for use (MacKenzie, 1996). Imagine that the vertical axis of a graph is a continuum of various degrees of uncertainty about a particular technology. The horizontal axis comprises various categories of actors. At the extreme left, characterized by high uncertainty, are those directly involved in production of the device and of the knowledge incorporated in it – insiders who know details about the device which others do not and who are therefore aware of its fallibility. At the extreme right of the graph are the complete outsiders, those totally extraneous to the institutions propounding the technology in question and/or oriented to an alternative technology: in the case just examined, the opponents of the Star Wars programme or the army officials supporting the use of other weapons. In the middle lies the 'certainty trough' occupied by those who are loyal to the institution in question but not directly involved in development of the device: army generals, business managers and politicians who take the device as they find it and use it – in practice or in rhetoric – in their work (Figure 5.2). This approach is an attempt to interpret the diverse attitudes towards a technological device according to the social and institutional roles of the actors involved, and it reflects those developed by SSK to explain differences in attitude towards paradigms and anomalies. In other words, its purpose is to answer the crucial

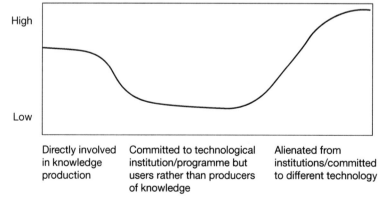

Figure 5.2 The certainty trough
Source: MacKenzie (1996: 256)

92 The sociology of technology

question in this area of study, which in relation to technology takes the form: 'Why can one man's efficient technological device be another man's questionable object?' (see Chapter 2 and Barnes, 1982b). Thus, the role of sociological explanation is more articulated than the SCOT approach, because it is not confined to interests in the strict sense, nor to the introduction phase of an innovation, but concerns itself with its subsequent use.

Finally, deduction itself may be subject to negotiation: the very concept of 'proof' may be understood differently by engineers, physicists, mathematicians and logicians. At the end of the 1980s, a chip named VIPER (Verifiable Processor for Enhanced Reliability) became commercially available. This was the first microprocessor chip whose reliability did not have to be tested inductively – i.e. by using it and seeing if any defects emerged (as usually happened with software and hardware products) – but it was given as mathematically proven, which was a feature of great importance for safety and security systems. In 1991, one of the companies that had purchased the licence brought a lawsuit against the British Ministry of Defence, which had described VIPER's design to be proven as a correct implementation of the specifications. Thus, mathematicians, logicians and information engineers found themselves in court arguing over what effectively counted as proof: the tons of computer printouts – the purported proof of VIPER's reliability consisted of seven million deductive inferences performed by another computer – or the understanding of those printouts by human beings? And how could one 'trust' the reliability of the computer which had performed the calculations proving that VIPER was perfectly reliable.[2]

Notes

1 For a critical reconsideration of White's thesis see, for instance, Hall (1996).
2 The trial was not concluded because the company that had brought the lawsuit went bankrupt (Mackenzie, 1993).

6 'Science wars'

1 Hoaxes and experiments

In 1989, more than 60 laboratories around the world officially announced that they had replicated Pons and Fleischmann's experiment and achieved 'cold' nuclear fusion. At least two Nobel prizes for physiology have been awarded for discoveries that subsequently proved non-existent: the one awarded in 1903 for the discovery of phototherapy, and the 1927 prize for the treatment of dementia paralytica. For years, the National Institutes of Health gave large amounts of 'ad personam' funding to the virologist Peter Duesberg, today regarded as 'a public menace' by broad sectors of medical research because of his heterodox views on the aetiology of AIDS.

What do these facts show? That physicists at numerous research institutions are bumbling incompetents, or that the criteria used to allocate large sums of research funding and prestigious awards should be revised? And what do Lacan, Baudrillard, Bergson, Feyerabend, Kuhn, Latour and Bloor have in common?

These questions are at the centre of a wide-ranging cultural debate which has recently involved sociologists of science as well. The debate was sparked, among other things, by the 'hoax/experiment' perpetrated in 1996 by Alan Sokal, a physicist at New York University. Sokal sent a paper entitled 'Transgressing the Frontiers: Towards a Hermeneutic Interpretation of Quantum Gravity' to the journal *Social Text*. The paper was unhesitatingly published by the journal, even though it was a mishmash of gibberish on physics and mathematics, simply because – according to Sokal – 'a) it sounded good and b) it flattered the editors' ideological preconceptions' (Sokal, 1996b: 62).

The article was, in fact, an entertaining parody of a certain academic style of writing, somewhat along the lines of the already-cited paper by Perec (Chapter 4). Sokal went further, however, insisting –

with wide coverage in the media – that his parody was an 'experiment' with great educational and 'political' value. What, therefore, were the hypotheses behind his experiment and what exactly did it prove? In numerous writings commenting on and justifying his hoax, Sokal has said that his intention was to pillory 'intellectual laziness and weak scholarship'. Eventually, Sokal declared that he wanted to 'combat a currently fashionable postmodernist/poststructuralist/ social-constructivist discourse – and more generally a penchant for subjectivism ... inimical to the values and future of the Left' (Sokal and Bricmont, 1997, English trans. 1998: 270). On the other hand, even Nobel physicist Steven Weinberg, one of Sokal's most enthusiastic supporters, admitted that he was unable 'to judge what (this experiment) proved' (Weinberg, 1996: 1).

Nor is any great help forthcoming from the book published by Sokal together with physicist Jean Bricmont with the intent of explaining and extending the scope of his assault on postmodernism (Sokal and Bricmont, 1997, English trans. 1998). In the course of this book, the declared critical targets range through:

a 'postmodernism' (p. 4);
b 'a radical version of postmodernism' (p. 183);
c 'social constructivism' (p. 269);
d 'epistemic relativism' (p. 50ff);
e 'the repeated abuse of concepts and terminology coming from mathematics and physics' (p. 18).

Yet, even if one accepts Sokal's personal reconstruction of postmodernism, it is difficult to see how Lacan can be included under that heading, even less so Feyerabend and Bergson. And what about Latour, who has devoted one of his best-known books, tellingly entitled *We Have Never Been Modern* (1991), to demonstrating that modernity has not been superseded – nor even, indeed, reached? And then what about the attribution of Kuhn to the postmodern? Although one is intrigued to learn from Sokal and Bricmont that there are '*two* Kuhns – a moderate Kuhn and his immoderate brother – jostling elbows throughout the pages of his immoderate brother' (p. 75, italics in the original). The confusion seemingly continues in each of the strands of postmodern thought examined, so that Latour is described as an 'exponent of the strong programme' (but here the error of chronology is a minor one, being a mere 15 years adrift). Oddly, not a single chapter in the book is devoted to Lyotard, one of the key representatives of postmodernist thought with his *The Postmodern Condition*.

The criticisms brought by Sokal and Bricmont against specific approaches largely use outworn arguments that other scholars have already (and better) deployed. Even the most intransigent positivist would be wary of attributing to the 'strong programme' the assertion that 'only social factors have an explanatory role', or of analysing Feyerabend's contribution on the basis of his provocative 'anything goes'. I presume it superfluous to point out that the transfer of metaphorical images among different scientific sectors is widely recognized as one of the main creative impulses in intellectual activity, or that it has been the frequent practice of such 'visible scientists' as Prigogine and Thom.

As to the value of Sokal's experiment, it is doubtful that *Social Text* is representative of all the categories that he attacks; it is certainly not so of sociology of science, nor of philosophy of science, which are Sokal's two main targets. In particular, if it was his aim to conduct a genuine experiment, he would have done better to select one of the many social science journals that use external referees (which *Social Text* does not).

An experiment/hoax of much greater significance than Sokal's was conducted in 1987 by William Epstein (Hilgartner, 1997). Epstein submitted an article in two different versions, but based on the same statistical data, to 147 social work journals: 74 of these journals received the article with a 'positive' conclusion (the social intervention had worked) and 72 received the one with the 'negative' conclusion. The first version was much more frequently accepted for publication, but when Epstein revealed his experiment, the reactions were much more violent than those provoked by Sokal's. It was even proposed that he should be struck off the professional register for using the journal editors as unwitting 'guinea pigs', and for breaching the principle of trust on which academic work is based.

It is not my purpose, however, to belittle Sokal's project, which certainly proves something, although that something may not have been among his original intentions. It shows, first, that processes like the selection of articles for publication is fraught with political, social and cultural elements. This is an aspect well known to sociologists of science: long before Epstein, in fact, Merton (who certainly cannot be accused of relativism, let alone constructivism) had pointed out in the 1960s that the presence of a Nobel prize-winner among the authors of an article could exponentially increase the likelihood that it would be accepted for publication (see Chapter 1).

Second, the Sokal case shows that the significance of an experiment depends on the 'scientific-cultural' context in which it takes place.

And here it should be stressed that lively debate on many of the issues raised by Sokal had already been in progress in the Anglo-Saxon countries at least since 1994, when Gross and Levitt's book *Higher Superstition – the Flight from Science and Reason* openly accused 'certain sectors of the American intellectual Left', and in particular historians and sociologists of science, of fomenting hostility to science. In other words, the commentators and journalists who seized so gleefully on Sokal's hoax acted no differently from the editors of *Social Text* when they uncritically accepted something that 'sounded good and flattered their ideological preconceptions'.

Finally, 'transgressing the frontiers' between the natural and social sciences is routine practice on both sides of the 'two cultures', and Sokal and Bricmont engage in it themselves when they set out to give lectures to humanists.

2 Have we never been sociologists of science?

Much more significant, and even more radical, is the critique carried forward within the sociology of science itself. We have already met Latour and his contention that the scientific fact should be considered, not as a peg on which to hang social factors but as the outcome of a complex network of alliances and translations (see Chapter 4). In developing his thesis, Latour proposes that modernity itself should be viewed as centred on a contradiction (Latour, 1991). On the one hand, in fact, modernity constantly creates 'hybrids' by mixing nature and culture. Suffice it to read the pages of any newspaper to find dozens of such hybrids: AIDS, the hole in the ozone layer and mad cow disease are all objects in which technical-scientific and social-political aspects are inextricably bound up with each other. On the other hand, modernity theorizes the separation and purging of the natural dimension from the human component. Over here are facts, microbes, missiles, prions; over there society, the worries of ecologists, the interests of the pharmaceutical companies, the intentions of heads of state. Over here stands Boyle, who saw consensus guaranteed by his vacuum pump, a non-human actor, an immutable fact 'whatever may happen elsewhere in theory, metaphysics, religion, politics or logic' (Latour, 1991, English trans. 1993: 18); over there stands Hobbes, for whom any agreement on knowledge that omitted the political dimension was impossible. The 'victory' of Boyle and his air pump made possible the formidable 'double game' of modernity: using the natural sciences to 'debunk the false pretensions of power and using the certainties of the human sciences to uncover the false pretensions of the natural sciences' (Latour, 1991, English trans. 1993: 36).

> Native Americans were not mistaken when they accused the Whites of having forked tongues. By separating the relations of political power from the relations of scientific reasoning while continuing to shore up power with reason and reason with power, the moderns have always had two irons in the fire.
> (Latour, 1991, English trans. 1993: 38)

The problem with much of the sociology of science, according to Latour, is that it has been duped by this 'double game' just as much as any other discipline. Trying to use society to explain science means accepting and reinforcing this separation, which is, itself, a hybrid of nature and culture. The strong programme did not fully apply the symmetry principle that it preached. One cannot be a constructivist with nature and a realist with society, using it as the bulwark for one's analyses of scientific practice. While the hole in the ozone layer is too social to be considered a purely natural fact, political strategies are too concerned with embryos and stem cells to be reduced to interests. We cannot consider Pasteur's bacteria without considering French society and politics of the 1800s, or Edison's electric light bulb without examining the American economy at the time. Nor can we consider the contemporary concept of family, or indeed the ideas of life and death, without taking account of assisted reproduction techniques or of the mapping of the human genome. Whence derives, according to Latour, the stagnation in which the sociology of science has languished in recent years.

3 What sociology of science?

Not surprisingly, the positions set out in the previous section have provoked strong criticism from the SSK.[1] While some of Latour's conclusions appear debatable, it is undeniable that since the early 1990s, the proliferation of case studies and increasing internal specialization of the field have not been matched by a corresponding growth of theory.

I shall devote the rest of this chapter to description of a line of inquiry that has hitherto received relatively little attention, but which will enable me to re-examine a number of key themes in sociology of science. To introduce this theory I shall refer to a short essay by

Barnes (1983). This is an extremely dense and abstract study, almost entirely devoid of examples and whose difficulty is probably one of the reasons why so few have taken up its challenge. I shall do my best to clarify Barnes' argument and to repay the effort required to read the next few pages.

The point of departure is the manner in which we name the objects of the world on the basis of two broad categories of terms. The first comprises so-called 'N-terms', which are applied by a process in which the empirical properties of an object are compared against a model (pattern-matching). For example, a person will have learnt, on the basis of examples that s/he has been shown or of what s/he has been told, those features of trees which characterize a certain entity as a 'tree'. But there is another and equally idealized way to assign properties to objects. This is based, not on the intrinsic properties of the object but on the way other people define it. A predicate of this type Barnes calls an 'S-term'. For example, 'female' can be considered an N-term in Barnes' sense because there are empirical features which distinguish the female from the male. 'Married' is, instead, an S-predicate because it can be applied only on the basis of what one has been told about the person concerned. By pronouncing two persons husband and wife, for example, the priest or the mayor makes them such. These types of predicates have the characteristic of making themselves 'true', so that, for instance, a person is a leader if a sufficient number of people recognize him or her to be such. Social institutions like marriage or money depend on forms of behaviour, thought and collective conversation which are performative and self-referential.

Imagine a 'designating machine' which performs highly routinized designation operations. The machine consists of two sections. Fed into the first section of the machine are details or characteristics of objects which are compared against already-existing patterns. The details that match these patterns are sorted into one category, those that do not are sorted into another. The second section designates as 'N' the details allocated to the first category, and as 'non-N' those allocated to the other.

Now imagine a machine which performs the same operation for S-terms. While the importance of the N-machine is obvious, the S-machine seems purely tautologous ('this is an S because I say it is an S'). The situation changes if we have a series of S-machines operating in parallel, as in a community. The more that certain details are designated S, the more S they become. For a person working with an S-machine, individual designations have inductive value: in other words, an individual social actor may in practice often treat

S-terms 'as if' they were N. If a Mafia gang is sufficiently large and stable, the boss's leadership is taken for granted. The solidity of a bank is assumed by its hundreds of thousands of customers, although it is they who render it solid by placing their trust in it, and by making deposits based on that trust. The case of money is exemplary: the self-referentiality that gives value to money has become so taken for granted that it is invisible (MacKenzie, 2001). The concept of 'self-fulfilling prophecy' has been familiar to sociologists for more than 50 years – and perhaps it is no coincidence that it was first formulated by a sociologist so closely interested in the sociology of scientific knowledge as Merton (1968b). Rumours of a bank's insolvency induce customers to withdraw their money, so that the bank does, indeed, become insolvent. The exclusion for many years of black workers from the American trade unions – based on the prejudice that they were unreliable if strikes were called and willing to work for low wages – increased the likelihood that they would work for low wages and would break strikes.

S-elements also play a role in scientific activity. Examples are the researcher who uses a bottle of hydrochloric acid because attached to it is the label 'hydrochloric acid'; the laboratory technician who considers the temperatures significant for a particular experiment to be the temperatures that others have deemed significant;[2] or the researchers studied by Merton and Collins, who adopted a certain attitude towards an experiment because it had been carried out by a researcher whom they regarded as 'reliable' or who worked at a particular institute. In this sense, a large part of the progress achieved by the social studies of science has taken the form of the discovery of S-aspects within N-terms (MacKenzie, 2001).

Let me give a more detailed example. None of us is able to 'recognize' an electron, not even J.J. Thomson and Millikan, who measured certain of its properties like mass and charge. Our idea of the electron, like Thomson and Millikan's ability to talk about the same thing without 'seeing' it, is therefore based on an indirect process of recognition, and on the application of a general model to a specific case (Bloor, 1995).

However, the fact that a model is considered to be such is, in itself, an S-type operation and therefore pertains to the social dimension. What gives a representation – for example, that of the atom as the solar system or the electron as a magnetic pole – the status of a model is the fact that scientists use it and consider it to be a model. The more researchers use the metaphor which likens the atom to the solar system, the more that metaphor becomes established and taken for granted: 'Something is a model only if a sufficient number of people treat it as a model, just as something is money only if a sufficient number of people treat it as money' (Bloor, 1995: 12).

As happens in the case of a standard technology or a software, the use of a more common model or a widely employed metaphor means that a wider range of 'materials', 'supplies' and exchanges with other users becomes available. The status of a model, the fact that it is considered appropriate or inappropriate, central or marginal, influences its application and its very evolution. Conflicting results and the discovery of the electron's undulatory properties did not affect its solidity and credibility *qua* institution; indeed, electrons were intelligible and framable on precisely that basis. The guarantee for the routine and unproblematic application of a concept is, therefore, its grounding on consensus, besides the fact that what we customarily take to be the meaning of a predicate is 'the institution of its use'. What is it that guarantees that Barnes' machines will continue correctly and coherently to apply the labels of 'fish' or 'electron'? How can we be sure that hundreds of machines, operating on thousands of examples, will not eventually come to classify a 'fish' as a 'non-fish', and vice versa? The contention of Barnes and Bloor is that the sociology of science can take the field at this point: guaranteeing the correct 'pattern-matching' by each machine is the fact that several machines are operating in reciprocal interaction. In the absence of this interaction, it does not even make sense to talk of the 'right' or 'wrong' application of Ns. Thus, the group as the basis of consensus becomes the condition for normativity and therefore for meaning (Bloor, 1996).

Our ability to address and know the natural world – an ability that the SSK in no way denies – therefore rests on a subtle social 'prop'. Only this support makes it possible for us to distinguish between the proper and improper applications of a concept, to organize our knowledge-gathering activities, and to focus only on certain elements, so that we are not constantly forced to reboot our understanding from scratch. The orientation of the strong programme can thus be reformulated as follows:

> Meaningful reference to an independent reality requires a social institution to make it possible ... one way to define a 'strong programme' in the sociology of knowledge is through the claim that all concepts have the character of institutions, or that all natural kind terms involve, as a necessary element, the self-referential machinery characteristic of social kind terms.
>
> (Bloor, 1995: 20)

'Science wars' 101

In my view, this recasting of the SSK displays a number of interesting features with respect to the approach in its original form. First, it clarifies that 'social' is not synonymous with 'context'. A social dimension and, therefore, a sociological analysis, is also applicable to communities of specialists or technicians (MacKenzie, 1996). Second, it offers a more systematic answer to some of the questions that we have seen preoccupy science studies from their beginnings, starting with the different attitudes of researchers or groups of researchers to the same empirical evidence. Kaufmann's observation that the mass of an electron increased with its velocity was interpreted by Thomson in 1901 as demonstrating the importance of ether, an 'institution' then losing credibility but still relatively well-established among English physicists (Bloor, 1995).

On this basis, MacKenzie's study of the proof of the four-colour conjecture adds another interesting example from mathematics (see Chapter 3). The hypothesis that four colours suffice to shade any geographical map in such a way that countries sharing a border are never of the same colour was first formulated in 1852. But it was only proved in 1976 by Appel and Haken with the help – almost unprecedented in mathematics at the time – of complex computerized procedures. Fierce debate was provoked by the proof. Numerous mathematicians refused to accept that a demonstration made by computer could actually count as a 'mathematical proof'[3] and argued that Appel and Haken's work, however interesting, was 'something else' (an experiment?). Others even suggested that the concept itself of mathematical proof would now have to be revised (MacKenzie, 1999). As Bloor describes it, the history of the changing definitions

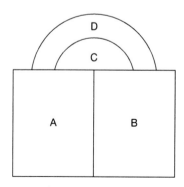

Figure 6.1 A map which cannot be coloured with fewer than four colours
Source: MacKenzie (1999: 11)

of a polyhedron does not tell us a great deal about the social dimension of science, apart from the fact that definitions and theorems are amply negotiable among scholars. From this new perspective, however, the social dimension appears to be some sort of precondition for scientific knowledge itself. The status of 'proof' or 'counter-example' given to an experimental result, or that of 'model' or 'experiment', contains S-type elements. Consider, for example, the well-known episode following the 1986 Challenger space shuttle disaster, when the simple manipulation of a rubber ring and ice water by the physicist Richard Feynman during a press conference was regarded as a genuine experiment and as conclusive evidence of the cause of the accident.[4]

At a more general level, by tying the cognitive and normative dimensions together, this recasting of the SSK yields a more systematic description of the interaction between knowledge processes and social processes – a description that was ill-defined and largely mechanistic in the original causal explanation based mainly on interests. The description of the social dimension as an interconnected web on which our knowledge of nature is deposited is not, in the final analysis, very different from that of actor-network theory; but it places greater emphasis on its social and cultural nature, while Latour and Callon gave a more 'political' representation of it.

Finally, whereas in its original formulation the SSK was based on a traditional sociological framework, the new approach seems able – especially in Barnes' work – to integrate various currents of inquiry: symbolic interactionism, Goffmanian microsociology and ethnomethodology. Moreover, this integration comes about at a theoretical level, and is not merely empirical, as it was when exclusive use was made of case studies.

However, this reformulation has not been immune to criticisms, some of which Bloor himself has rebutted by declaring that none of the features described above imply that 'all predicates are S-predicates but ... that all predicates have S-type features or aspects to them [and] do their work in virtue of having a self-referential component in their use. They work in virtue of being social institutions' (Bloor, 1995: 6). Thus, a fundamental point that the critics of the sociological approach to scientific knowledge have often failed to grasp is made even more explicit. In this sense, the approach of Barnes and Bloor is diametrically opposed to that of Collins and other proponents of the 'methodological relativism' whereby every discourse must be treated as self-referential even if it refers to 'non-social' reality. The endeavour of Barnes and Bloor is not to expand the social component but to do the opposite: eliminate it entirely, in order to show that the social is a thin yet essential layer. If the institutions – i.e. that part of reality created by the act of referring to it – are removed, then any possibility of shared meaning collapses (Bloor, 1996).

Thus, accepting that knowledge has a social dimension is not to question the value of that knowledge. Social processes do not impede or attenuate our ability to know the outside world: on the contrary, such knowledge is made possible precisely by the presence of social institutions and of what we call society.

Bloor also seeks to forestall an excessively radical Durkheimian reading whereby the discourse of scientists – for instance that of Thomson and Millikan about the electron – is nothing but a social product, a way of talking about society entirely akin to the way in which the divinity was talked about in primitive societies. An electron is not at all an entirely self-referential term, unlike what one of social type might be. Simply, 'the involvement of Thomson and Millikan with the natural, non-social world was also, and simultaneously, involvement with other people, through a shared culture' (Bloor, 1995: 20).

Note that this recasting of the SSK is also a reply to one of the questions raised by Kuhn: how do scientists immersed in incompatible paradigms manage to 'communicate'? And consequently, how is scientific change possible? One plausible – and only apparently simplistic – answer is that different scientific sub-cultures belong to the same culture and to the same society, and it is this that enables them to understand each other.[5] In Fleck's terminology, a researcher simultaneously belongs to several thought-collectives: that of his/her specific sector of inquiry, but also that of his/her religious affiliation, or that of the political party which s/he endorses, besides that of the more general collective of the society and the culture in which s/he lives. And it is precisely in the exchanges and intersections among these thought-styles that the most significant changes in scientific knowledge occur (Kuhn's revolutions) (Fleck, 1935).

More detailed examination of the foregoing proposal would probably reveal further flaws. Here, however, I shall conclude with a point central to numerous criticisms. I refer to the role of the 'macrosocial' category of interests to which the scholars of the strong programme initially gave such importance. What has happened to interests in the new framework described by Barnes and Bloor?

To tell the truth, it is difficult to say. While on the one hand interests seem to be fragmented in a dimension which stresses routine adherence to practices and institutions of shared meaning, on the other it seems that Bloor, especially, is unwilling to relinquish them. By choosing to adhere to one model-institution rather than another, a scientist seeks to maximize his/her opportunities to reduce costs.

I shall not discuss this position in detail. In any case, it seems to suffer from certain of the shortcomings of an excessively rationalist approach centred on the scientific actor. General criticisms of the view of institutions as 'rational solutions' to the problem of minimizing transaction costs are easy to find in the sociological literature.[6] Instead, perhaps surprising is the absence of any connection with such sociological concepts as identity. Just as a worker may take part in a protest demonstration contrary to his/her strictly utilitarian convenience, because self-recognition as a member of a certain group is the precondition for rational choice itself (Pizzorno, 1986), so for a scientist the use of instruments, procedures and concepts shared with his/her colleagues is a precondition for the assessment of empirical results or theoretical hypotheses. Thus, for a researcher, adherence to tradition in certain contexts – for example that of the Neapolitan mathematicians belonging to the 'synthetic' school (see Chapter 3) – may be a key element in his or her identity as a scientist. In other situations, innovation and the superseding of traditional models, or more recently the ability to produce 'patentable' and 'marketable' knowledge, may be equally crucial to identity-forming. The use of mathematical-statistical models – which in the past was extraneous to the identity of researchers in the biological disciplines – is today indispensable. The use of computers has become part of the identity of mathematicians. Consider, likewise, the renewed importance that the concept of 'trust' assumed in the above-cited cases of contemporary research in mathematics or physics.[7] When complex mathematical proofs or experiments on subatomic particles require months of calculations and equipment available at only a handful of research centres, a large part of the scientific community is forced to delegate control over its results to an extremely small number of colleagues. This inevitably reinforces the mechanisms whereby the validity of a result depends on the visibility, reputation and institutional position of the researcher who has produced it, as pointed out by both Merton and Collins and Pinch.[8]

It may be here that Science and Technology Studies shake off, at least to some extent, the legacy from Merton's rejection of institutional sociology that induced their relative isolation from general sociological theory and their preference for linkages with other disciplinary sectors. But the principal merit of this recasting of SSK is,

I repeat, its greater emphasis on an aspect of the sociology of science which clears up a misunderstanding that has misled most critics, as well as the protagonists themselves.

> This idea of competition between what is logical and natural on the one hand, and what derives from culture and society on the other, is deeply entrenched. [According to this idea] classifications may conform to the objective facts of nature *or* to cultural requirements. They may be logical or social. But it is the very opposite of what careful examination reveals: we need to think in terms of symbiosis, not competition.
> (Barnes, 1982a: 197)

Thus, Merton's analysis of the 'self-fulfilling prophecy' strikes Barnes as incomplete. This is not merely a pathological feature: a bank considered to be solid is no less a self-fulfilling prophecy than an insolvent bank. The vicious circle and the virtuous circle that sustain our everyday routines are two sides of the same coin.

I leave the final word to Ludwik Fleck, medical doctor and pioneer in the sociology of knowledge, who once again got the point before anyone else:

> those who consider social dependence a necessary evil and an unfortunate human inadequacy which ought to be overcome fail to realize that without social conditioning no cognition is even possible. Indeed, the very word 'cognition' acquires meaning only in connection with a thought collective.
> (Fleck, 1935, English trans. 1979: 43)

Notes

1 See e.g. Bloor (1999).
2 Hacking (1992) similarly describes three types of conditions ensuring stability in science practice: anachronism (doing different things and accepting 'on faith most knowledge derived from the past'); the presence of several different strands (a break in a theoretical tradition does not necessarily mean a break in the use of experimental instruments); the turning of various elements into 'black boxes' (e.g. 'statistical techniques for assessing probable error, ... standard pieces of apparatus bought from an instrument company or borrowed from a lab next door') incorporating 'a great deal of preestablished knowledge which is implicit in the outcome of the experiment' (Hacking, 1992: 42).
3 'We cannot possibly achieve what I regard as the essential element of a proof – our own personal understanding – if part of the argument is hidden

106 *'Science wars'*

   away in a box' was one of the objections raised by mathematicians during the debate on Appel and Haken's results. Another mathematical proof achieved by means of computerized calculations was Lan, Thiel and Swiercz's demonstration of the non-existence of finite projective planes of order ten, obtained by examining 1,014 cases after thousands of hours of calculations on one of the most powerful computers then available, the Cray-1 at the Institute for Defense Analysis of Princeton (MacKenzie, 1999).
4 Gieryn and Figert (1990). On 11 February 1986, the physicist and Nobel prize-winner Richard Feynman, a member of the presidential commission investigating the Challenger explosion, told a press conference that he could demonstrate the cause of the accident. He took a piece of the rubber ring used to prevent the escape of hot gas from the join between the segments of the rocket. He immersed it in a glass of ice water, squeezed it in a clamp, and held it up to show that it could not spring back to its original shape. The material's scant reactivity at low temperatures (like that on the morning of the launch) had caused the Challenger disaster.
5 See on this also the last writings of Feyerabend (1996a, 1996b).
6 For wide-ranging discussion see e.g. March and Olsen (1989), Powell and DiMaggio (1991).
7 For an analysis of the concept of trust in general sociological theory see e.g. Coleman (1992, Chapters 5 and 8).
8 According to one of the mathematicians interviewed by MacKenzie within the framework of the Four Colour Conjecture case, many of his papers had been accepted 'surprisingly quickly' by specialist journals, making him suspicious that the referees had looked 'only at the author and the theorem, without examining the details of the putative proof' (MacKenzie, 1996: 261, n. 39).

7 Communicating science

An article on the front page of a newspaper describing a successful experiment to clone a sheep, a TV weatherman talking about variations in atmospheric pressure for the next few hours, or a science museum where visitors can make experiments to understand the principles of gravity: these are some of the many and diverse situations in which 'laymen' – non-scientists – come into contact with science. What impact do they have on the image and public perception of research? And what bearing do they have on scientific activity itself?

Both scientists and scholars of scientific activity – including sociologists of science – have often dismissed situations such as these as having little effect on the understanding of science. In recent years especially, the theme of the public communication of science – or the 'popularization of science', to use a widespread albeit unsatisfactory expression – has gained greater importance and visibility. Indeed, complaints are often voiced about the public's low level of 'scientific literacy', with calls being made for the more vigorous dissemination of scientific knowledge – an objective by now on the agenda of numerous national and international public institutions.[1]

1 The mass media as a 'dirty mirror' of science

Scientific communication addressed to the layman has a long tradition. Consider the numerous popular science books written in the eighteenth century to satisfy growing public interest, especially among women, of which instances are Algarotti's *Newtonianism for Ladies* or de Lalande's *L'Astronomie des Dames*, the numerous accounts of scientific discoveries published in the daily press, or the great exhibitions and fairs that showed visitors the latest marvels of science and technology (Raichvarg and Jacques, 1991).

108 Communicating science

However, communication practices in science have developed mainly in relation to two broad processes: the institutionalization of research as a profession with higher social status and increasing specialization; the growth and spread of the mass media.

The idea that science is 'too complicated' for the general public to understand became established as a result of advances made in physics during the early decades of the 1900s. In December 1919, when observations made by astronomers during a solar eclipse confirmed Einstein's general theory of relativity, the *New York Times* gave much prominence to a remark attributed to Einstein himself: 'At most, only a dozen people in the world can understand my theory'.[2]

This idea underpins a widespread conception, if not an outright 'ideology', of the public communication of science. The other cornerstones of the conception are the need for mediation between scientists and the general public (made necessary by the complexity of scientific notions), the singling out of a category of professionals and institutions to perform this mediation (scientific journalists and, more generally, science communicators, museums and science centres), and description of this mediation by means of the metaphor of translation. Finally, it is taken for granted that a wider diffusion of scientific knowledge requires greater public appreciation of, and support for, research (Lewenstein, 1992b).

This 'diffusionist' conception, indubitably simplistic and idealized, which holds that scientific facts need only be transported from a specialist context to a popular one, is rooted in the professional ideologies of two of the categories of actors involved. On the one hand, it legitimates the social and professional role of the 'mediators' – popularizers, and scientific journalists in particular – who undoubtedly comprise the most visible and the most closely studied component of

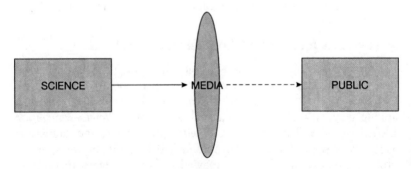

Figure 7.1 The traditional conception of the public communication of science

the mediation. On the other hand, it authorizes scientists to proclaim themselves extraneous to the process of public communication so that they may be free to criticize errors and excesses – especially in terms of distortion and sensationalism. There has thus arisen a view of the media as a 'dirty mirror' held up to science, an opaque lens unable adequately to reflect and filter scientific facts.[3]

2 Journalists and the difficult art of mediation

Research has long concerned itself with the media coverage of science, and with the public for whom such coverage is intended. Studies on the matter have typically examined the representation of scientific topics by the media, for example by asking one or more scientists to appraise the quality of the journalistic treatment given to a particular issue. The results have usually led to calls for greater accuracy, for closer interaction between journalists and specialist sources and, in general, for efforts to mimimize the elements that cause 'disturbance' in communication between scientists and the general public, which otherwise would be straightforward.

Researchers have also pointed out the tendency for the media – mainly the press, given that very few systematic studies have been conducted on coverage by television or radio – to over-represent certain disciplinary areas (biomedicine for example); to depend on specific events or on social rather than scientific priorities; and to emphasize risk over other features. For example, in a long-period analysis of the coverage by the American popular press of infectious diseases like diphtheria, typhoid and syphilis, Ziporyn has shown the greater importance of social values – rather than scientific discoveries – in determining the nature of such coverage (Ziporyn, 1988). It is rather rare, for example, for a mathematical discovery to be reported on the front pages of the newspapers or by prime-time news bulletins. The selection of scientific themes or news stories is often conditioned by the occurrence of 'newsworthy' events or by the possibility to link them with other topics of a non-scientific nature. The prominence given to the 'mad cow' emergency in Italy – well before cases were discovered in the country and after 11 years of crisis in Britain – was not unrelated to the importance attributed to the theme of European integration at the time. In 1997, the announcement of the birth of the cloned sheep 'Dolly' was given blanket coverage for almost a month by a press already very aware of themes like embryos, *in vitro* fertilization and abortion, while the announcement made four years previously of a significant advance in human cloning had been ignored.⁴ The 'scientific experts' selected by the mass media to comment upon a specific issue are not necessarily the ones best qualified to do so: more important in the choice of an expert by journalists may be his/her visibility externally to the research community (as the member of an advisory committee, as a politician, as a popularizer), the fact that s/he is also interesting from a human point of view or that s/he is willing to talk about a wide range of topics, and that his/her use can be easily justified (because s/he belongs to a particularly prestigious institution or has received particular awards or honours) (Goodell, 1977; Peters, 2000).

However, long-period analysis of the treatment of scientific themes by the non-specialist press shows that it presents scientific activity as largely 'progressive', as beneficial to society, and as consensual. Such coverage is found to adhere closely to specialist sources – often cited directly or indirectly – and indeed in linguistic terms is not particularly distant from specialist communication.⁵ Numerous studies have reported that science journalists are increasingly inclined to believe that a scientific background is essential for their work, and consider their profession a means to bolster the image and importance of science vis-à-vis public opinion. From this point of view, one notes a relatively clear-cut distinction between scientific journalists – those who deal with science on a full-time basis, writing for specialist newspaper sections or popular science publications – and 'general news' journalists who may on occasion find themselves dealing with scientific topics. As regards professional values, the former stand much closer to the scientific community than to the general public: they more often view their 'professional mission' in terms of popularization, when not of education and cultural edification. News journalists by contrast see it as their duty to express public concerns and demands: they describe their mission in terms of public opinion's need for information, which justifies their indifference to the priorities set by the scientific agenda.⁶

3 Is the public scientifically illiterate?

The diffusionist – pedagogical-paternalistic – conception of the communication of science has long informed studies on public scientific knowledge as well. First conducted in the US during the 1950s, research on interest in science and scientific information among the general public and its awareness of science has, since the 1980s, become common in numerous countries. The results of this research have frequently been used to decry the public's scant interest in science, and its excessively low level of 'scientific literacy', and to call for quantitative and qualitative improvements in scientific communication addressed to the public at large.[7] Although a certain degree of public ignorance is undeniable – for instance, European surveys on the public perception of biotechnology found that more than 30 per cent of the population thought that, unlike genetically modified tomatoes, 'normal' ones do not contain genes[8] – numerous criticisms have been made of this approach. The indicators used to measure the public understanding of science are often debatable. For example, in 1991 a study by the National Science Foundation complained that only 6 per cent of interviewees were able to give a scientifically correct answer to a question on the causes of acid rain; but it neglected the fact that specialists themselves still disagree as to what those causes actually are. Other studies have emphasized the complex articulation of public images of science, where a belief that astrology is a scientific discipline – classified by numerous surveys as indicative of scientific illiteracy – is often accompanied by a sophisticated understanding of science.[9] Anything but established, moreover, is the linkage among exposure to scientific information in the media, level of knowledge, and a favourable attitude towards research. As regards biotechnology, for example, recent studies have highlighted substantial levels of scepticism and suspicion even in the best-informed sectors of the population.[10]

More generally, the cleavage between expert and lay knowledge cannot be reduced to what the 'deficit model' of the public awareness of risk regards as merely an information gap between specialists and the general public. Factual knowledge is only one ingredient of lay knowledge, in which other elements (value judgements, trust in the scientific institutions) inevitably interweave to form a complex which is no less articulated than the expert one. The source which Europeans regard as providing the most trustworthy information about biotechnology, for example, are consumer associations (Gaskell *et al.*, 2000). Scientific information may be ignored by the public as irrelevant or scarcely applicable to their everyday concerns, as has been the case of information campaigns on what to do in the case of emergencies in communities located close to nuclear power plants. The representation of risk by medical experts, for instance, and the relationship between causes and effects in contemporary medicine, are increasingly expressed in formal and probabilistic terms. Yet, the perception of non-experts is inevitably based on subjective experiences and concrete examples. In a study carried out on English mothers who had refused to have their babies vaccinated as required by law, New and Senior discovered that this refusal had nothing to do with misinformation or irrational decision-making but instead sprang from a rationality at odds with that of medical experts. Many of the women interviewed, in fact, said that they personally knew other mothers whose babies had suffered serious disorders following vaccination, and that they had seen collateral effects in their own children (Lupton, 1995).

A classic example of the gap between expert and lay knowledge is provided by Brian Wynne's study of the 'radioactive sheep' crisis which erupted in certain areas of Britain at the time of the Chernobyl nuclear plant disaster in Russia. For a long time, Government experts minimized the risk that sheep flocks in Cumberland had been contaminated by radiation. However, their assessments proved to be wrong and had to be drastically revised, with the result that the slaughter and sale of sheep was banned in the area for two years. The farmers for their part had been worried from the outset, because they had direct knowledge based on everyday experience (which the scientific experts sent to the area by the government obviously did not possess) of the terrain, of water run-off and of how the ground could have absorbed the radioactivity and transferred it to plant roots. This clash between the abstract and formalized estimates of the experts and the perception of risk by the farmers caused a loss of confidence by the latter in the government experts and their conviction that official assessments were vitiated by the government's desire to 'hush up' the affair (Wynne, 1989).

According to some scholars, experts themselves reinforce the representation of the public as 'ignorant'. During a study on communication between doctors and patients in a large Canadian hospital, a questionnaire was administered in order to assess the patients' level of medical knowledge. At the same time the doctors were asked to estimate the same knowledge for each patient. The three main results obtained were decidedly surprising. While the patients proved to be reasonably well-informed (providing an average of 75.8 per cent of correct answers to the questions asked of them), less than half the doctors were able to estimate the knowledge of their patients accurately. This estimate was, in any case, not utilized by the doctors to adjust their communication style to the information level that they attributed to the patients. In other words, the fact that a doctor realized that a patient found it difficult to understand medical questions or terms did not induce him/her to modify his/her explanatory manner to any significant extent. The patients' lack of knowledge – the authors of the study somewhat drastically conclude – appeared in many cases to be a self-fulfilling prophecy, for it was the doctors who, by considering the patients to be ignorant and making no attempt to make themselves understood, rendered them effectively ignorant (Seagall and Roberts, 1980).

The diffusionist and linear conception of scientific communication is also highlighted by the scant attention paid to the influence of the images of science and scientists purveyed externally to information contexts and, particularly, in fiction. Yet, the few studies conducted on this topic show that these images are often of considerable importance in shaping the public perception of science and its exponents. Consider the role of works of fiction in sensitizing public opinion to AIDS or environmental risk. In the mid-1990s, the hereditary origin of breast cancer and preventive mastectomies was given particular salience by the British media, despite the low incidence of cases, because of the treatment given to the subject by a popular soap opera set in a hospital (Henderson and Kitzinger, 1999).[11]

4 The role of scientists

And what about scientists? Are they truly extraneous to these processes, passively at the mercy of the discursive practices of journalists and the incomprehension of the public?

Studies on the public communication of science tell us that they are not. For example, around 80 per cent of French researchers report that they have had some experience of popularizing science through the mass media.[12] Almost one fifth of the articles on science and medicine published in the last 50 years by the Italian daily newspaper *Il Corriere della Sera* have been written by researchers or doctors (Bucchi and Mazzolini, 2003). According to a broad survey of British scientists and journalists, more than 25 per cent of the articles on science that appear in the press start from initiatives – press releases, announcements of discoveries, interviews – by researchers and their institutions (Hansen, 1992). Moreover, researchers are often among the most assiduous users of science coverage by the media, on which they draw to select among the enormous mass of publications and research studies in circulation. A paper published in the prestigious *New England Journal of Medicine* is three times more likely to be cited in the scientific literature if it has first been mentioned by the *New York Times* (Phillips, 1991). The overall judgement passed by scientists on the media coverage of science – which as we have seen is markedly negative – becomes distinctly more positive at the analytical level when the quality of the media coverage on a specific issue is examined (Hansen, 1992). Finally, it is worth noting that the visibility of scientists in the media tends to display a pyramidal structure very similar to that of the distribution of other resources and remunerations in the scientific community. At the top of the pyramid stand a very small number of 'celebrities' who are frequently consulted on non-scientific issues as well – Nobel prize-winners being a typical example – and below them a broad base with very sporadic visibility (Goodel, 1977, 1987). These results have also prompted sociologists of science to interest themselves in the public communication of science, a topic on which their contribution has long been marginal with respect to other disciplines like social psychology, linguistics and media studies. This lack of interest in the public presentation and awareness of science can be explained by considering sociologists of science to be the most sophisticated victims of the traditional conception. As long as the public communication of science was considered a practice entirely detached from science, it was of scant relevance for those interested in the influence of social factors on scientific activity.

5 The public communication of science as the continuation of the scientific debate with other means

Science studies are highly critical of the traditional conception of the public communication of science. Instead of the sharp distinction between science and its popularization, they propose a 'continuity' model of scientific communication.[13] Along the continuum thus envisaged, differences – albeit only gradual ones – can be discerned among the diverse contexts and styles of communication/reception that inevitably exist in the expounding of scientific ideas.

One of the most detailed models of this *continuum* has been developed by Cloître and Shinn (1985) who identify four main stages in the process of scientific communication:

1. *Intraspecialist Level.* This is the most distinctively esoteric level, as typified by the paper published in a specialized scientific journal. Empirical data, references to experimental work and graphics predominate.
2. *Interspecialist Level.* Pertaining to this level are various kinds of texts, from interdisciplinary articles published in 'bridge journals' like *Nature* and *Science* to papers given at meetings of researchers belonging to the same discipline but working in different areas.

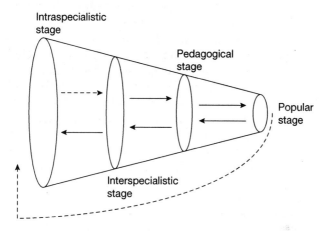

Figure 7.2 A model of science communication as a continuum
Source: Bucchi (1998), Shinn and Whitley (1985) and Hilgartner (1990)

3 *Pedagogical Level*. This is the level that Fleck calls 'textbook science', where the theoretical corpus is already developed and consolidated and the current paradigm is presented as complete (Fleck, 1935, English trans. 1979). The emphasis is on the historical perspective and on the cumulative nature of the scientific endeavour.
4 *Popular Level*. Cloître and Shinn include under this heading both articles on science published in the daily press and the 'amateur science' of television documentaries. They point to a larger quantity of metaphorical images in these texts and their marked attention to issues concerning health, technology and the economy.

A typology of this kind presents science communication as a continuity of texts with differences in degree and not in kind across levels. It invites us to imagine a sort of 'trajectory' for scientific ideas which leads from the intraspecialist expository context to the popular one, passing through the intermediate levels. This is a trajectory congruent with the theories that we have already met, from Fleck to Latour, on the construction of scientific facts. We may take as an example the tortuous process studied by Fleck (see Chapter 2), which led from a vague popular idea of 'syphilitic blood' to introduction of the Wassermann reaction and definition of the clinical distinctiveness of syphilis. This highly provisional definition, hedged about by doubts and methodological caveats (the first applications of the Wassermann yielded only 15 to 20 per cent of positive results when used on diseased subjects[14]), rapidly became an incontrovertible certainty in the eyes of the general public. Fleck used this example to reflect on the path followed by a medical-scientific notion from what he called the esoteric circle (the specialist community) to the exoteric one (the general public). Fleck compared a report on a clinical examination drawn up by one specialist for another with a report instead prepared for a general practitioner. Already at this point along the path, the report aimed at the general practitioner 'does not represent the knowledge of the expert. It is vivid, simplified and apodictic' (Fleck, 1935, English trans. 1979: 113).

Specialist exposition – the 'science of the journals' – is provisional and tentative. But when a theory makes its entry into the manuals it partly loses these features and is presented to the reader as generally accepted by the medical-scientific community: in other words, it becomes a 'fact'. A further step comes with the exposition characteristic of popular science; here 'the fact becomes incarnated as an immediately perceptible object of reality' (ibid.: 125). At the popular level, doubts and disclaimers disappear: the distinctions and nuances of specialist knowledge condense into elementary and compact formulas: AIDS is HIV, psychoanalysis studies 'complexes', the neurological theory that hypothesizes a division of tasks between the two hemispheres of the brain is transformed into a sharp antithesis between 'right-dominated' and 'left-dominated' people. The communicative path from specialist to popular science can thus be illustrated as a 'funnel' which removes subtleties and shades of meaning from the knowledge that passes through it, reducing it to simple facts attributed with certainty and incontrovertibility. Fleck stresses that this progressive solidification of knowledge then exerts an influence on specialists themselves.

> Certainty, simplicity, vividness originate in popular knowledge. That is where the expert obtains his faith in this triad as the ideal of knowledge. Owing to simplification, vividness and absolute certainty [popular knowledge] appears secure, more rounded and more firmly joined together.
>
> (ibid.: 113, 115)

The passage of a scientific notion through these various levels therefore cannot be described as the simple translation of an object from one communicative context to another. Each step – and this is one of the central messages of Fleck's book – involves a change in the notion. By way of analogy, something similar happens to characters and stories in literature. For example, none of Arthur Conan Doyle's original works contain the expression 'Elementary, my dear Watson'. Only after its introduction in a theatre production of the detective's adventures did the phrase come to epitomize Sherlock Holmes in the popular imagination.

Taking this intuition to its extreme, studies by sociologists of science based on the continuity model consider the level of popular communication to be the final (and often decisive) stage in the process of stylization, 'distancing from the research front', and production of factuality and incontrovertible truth which constructs scientific evidence (Collins, 1987).

> The more removed the context of research is from the context of reception in terms of language, intellectual prestige and skill levels, the easier it is to present their work as certain, decontextualized from the conditions of its production, and authoritative.
> (Whitley, 1985: 13)

The model is depicted by Figure 7.2 in the shape of a funnel, the purpose being to emphasize the growing solidity and simplification acquired by a scientific fact, level after level, until it becomes like a ship in a bottle: to be admired for its perfection but impossible to relate to its original components.

The continuity model can be considered a useful frame of reference insofar as it describes some sort of ideal flow of communication in routine circumstances. However, in some cases the level of public communication seems able to perform a more sophisticated role. An example is provided by the case of sickle-cell anaemia, which is a particular form of anaemia caused by a genetic deficiency in haemoglobin that causes the cells affected to assume an irregular shape. It afflicts only black people (in the US one black child in every 50 suffers from the disorder) and it is transmitted by heredity. The disease was first diagnosed by the physician James Herrick in Chicago. In 1949, Pauling demonstrated that sickle-shaped haemoglobin has a molecular structure different from the normal one; in 1957 the differences between the two molecules were defined; and in 1966 Marayama produced a complete model of the disease. However, the medical textbooks made no reference to sickle-cell anaemia until the mid-1970s, when it began to attract increasing public attention. After a series of television documentaries, funds were collected to help sufferers, and mention of sickle-cell anaemia was even made by President Nixon in a speech to the nation on health matters (February 1971). In 1972, funding for research on the disease was increased from one million dollars to ten million, and mass screening was organized throughout the country. This broad resonance with the public led to the inclusion of this form of anaemia as a topic of some importance in the medical textbooks (Balmer, 1990).

In this case we can speak of a 'deviation' to the public level, because the discourse did not follow the usual trajectory but passed directly at the public level, to then influence specialist ones. For example, the importance of appealing to the public in particular cases of change of controversy or paradigm has been variously hypothesized and studied (Jacobi, 1987). The wide and enthusiastic coverage given in 1919 by the daily press to the solar eclipse observations as confirming Einstein's theory of relativity – the *Times'* headline was 'Revolution in Science: New Theory of the Universe: Newtonian Ideas Overthrown' – played a crucial role in publicly settling an issue that was still being debated within specialist circles (Miller and Gregory, 1998).

Some conflicts – or more generally crises – seem impossible to resolve within the scientific community and must, therefore, be deviated to the public level.

Mention has already been made of how scientists make use of the information and images that circulate at the public level. Cloître and Shinn document how specialists appropriated a metaphor ('the ant in the labyrinth') originally used by popular science texts to explain the Brownian motion of particles (Cloître and Shinn, 1986). Around one third of the scholars involved in the debate on whether or not the mass extinction of the dinosaurs was due to the collision of the Earth with a meteor – another controversy with broad public resonance – stated that they had heard of Alvarez's impact hypothesis from the mass media (Clemens, 1994). The metaphor of the 'hole in the ozone layer', with its enormous impact on the media and public opinion, produced consensus at the public level at least one year before scientific consensus – extremely uncertain and controversial at the time – was reached on the effect of CFC on the atmosphere. Only subsequently was the metaphor re-imported into the specialist literature (Grundmann and Cavaillè, 2000).

It has, indeed, been argued that scientific discourse at the public level may in some cases resemble certain forms of political discourse in that it is only apparently 'public'. It is not really addressed to the public but is instead intended to reach a large number of colleagues rapidly. To do so, it uses the public level as a shared 'arena' where it is not necessary to comply with the constraints of specialist communication.[15] This prerogative of the public level is particularly important when communication must pass through several disciplinary sectors (a case in point being the hypothesis on the extinction of the dinosaurs, which concerned palaeontologists, geologists and statisticians) or several categories of actors. In the case cited of Pasteur's struggle to legitimize the anthrax vaccine and, more generally, the idea that diseases could be prevented by appropriate inoculation with the infectious agent, physiologists, doctors, veterinarians and farmers had simultaneously to be addressed. This difficult task was achieved by means of a public experiment organized in 1881 on a farm, where vaccinated and non-vaccinated cattle were infected with anthrax before the eyes of hundreds of people – including French and foreign newspaper reporters who wrote numerous detailed articles on Pasteur's success. Communication at the public level enabled the French physiologist to underplay still unclear theoretical issues by emphasizing practical ones – of great importance to some groups in his audience, e.g. farmers and politicians – such as the effectiveness and cheapness of his method. Moreover, immunization and the related practice of inoculation had long been familiar to the lay peasant culture (Bucchi, 1997). In 1919, Einstein was able simultaneously to address different disciplinary audiences (physicists, astronomers, mathematicians) through the popular press by giving interviews and writing articles on his theory of relativity (Gregory and Miller, 1998).

More recently, scientists who argued that the depletion of the ozone layer was due to CFC found the widely publicized image of the ozone 'hole' to be an effective device with which to alert researchers, politicians, environmentalists and public opinion to the emergency. The rapid public consensus achieved with the Montreal Protocol of 1987 – which provided for international agreements to reduce the CFC emissions responsible for ozone depletion – indirectly reinforced the status of a body of knowledge that was still being carefully debated by specialists.

Or again, when a new sector of research is being established or consolidated – as happened with climate studies, for instance, or the neurosciences in past decades – the public arena is vital if researchers are to communicate among different disciplines. In this way, communicating in public enables scientists not only to talk – albeit indirectly – among themselves (as Fleck pointed out) but also to gain recognition and construct a shared identity in terms of research interests and methods, thereby laying the basis for institutionalization of their sector.

In cases of 'deviation', therefore, the science communication process should be depicted as much more complex. For in these situations the public discourse of science does not receive simply what is filtered through previous levels but may instead find itself at the centre of the dynamics of scientific production. By and large, when talking about the public communication of science we are referring to at least two different things:

1 A 'routine' trajectory, consensual and non-problematic, which is adequately described by the continuity model. Despite its ideological connotations, 'popularization' is a sufficiently appropriate term for this process.
2 An alternative trajectory, which is the one represented by deviation to the public level, so that public communication acquires even greater salience and a more articulated role compared to specialist debate.

There are major formal and substantial differences between these two trajectories. At a formal level, when the popularization mode is activated, scientific problems are more frequently addressed in settings devoted explicitly to the communication of science: popular science magazines and the scientific pages of newspapers. Placing scientific notions in these media 'frames' gives them legitimacy and enhances their credibility. The most obvious example is the museum medium: the display of a scientific artefact in a museum tends automatically to confer the status of incontrovertible 'fact' upon it.[16]

On the other hand, when deviation occurs, scientific problems more frequently appear in generic media settings as well, like the news sections of newspapers and television newscasts.

At a more substantial level, in the case of popularization the outcome of communication at the popular level is relatively straightforward. As largely 'celebratory' (Curtis, 1994) discourse, popularization reinforces the certainty and solidity of theories and results. It is this process that the 'funnel' model of continuity depicts. When deviation processes instead occur, the outcome of communication at the public level cannot be determined a priori. For example, scientists increasingly use press conferences and newspaper articles to announce their discoveries. A certain period of time elapses before an article is published in a scientific journal (with a consequently greater risk that someone else will get into print first), and the anonymous examination of manuscripts by colleagues before publication prompts fears of plagiarism. In these cases, deviations to the public level can considerably accelerate the peer review procedure, although they may be viewed by colleagues as attempts to leap-frog the process and gain improper recognition outside the scientific community.

At this level, scientific facts (as well as the networks of professional and institutional actors surrounding them) may be consolidated, as the continuity model envisages, but they may also be dissolved, deconstructed or simply manipulated by social groups for their own purposes. The funnel does not necessarily taper off; it may expand again towards the specialist levels.

Social actors unrelated to the research community, like activists or the representatives of patients' associations may, in these situations, play a significant role in the definition of scientific facts.[17] Consider the case of research on AIDS, where drug testing procedures and the term itself for the disease were negotiated with groups of activists and patients' associations.[18] In the mid-1980s, AIDS patients participating in clinical trials of AZT (a drug that at the time was a promising candidate as a cure for the disease) developed remarkable technical competence that enabled them to substantially shape the trial procedure itself – for instance, by learning to recognize placebos and refusing to take them – and eventually to accelerate the FDA[19] standard authorization process. The testing of another drug for the treatment of an AIDS-related disease, *Pneumocystis Carinii Pneumonia* (PCP), aerosolized pentamidine, was performed by activist groups themselves after refusal by scientists to do so; the drug was approved in 1989 by the FDA on the evidence of only community-based research (Epstein, 1995).

Study of public scientific discourse in cases of deviation enables account to be taken of the 'plurality of the sites for the making and reproduction of scientific knowledge' (Cooter and Pumfrey, 1994: 254), and it also gives a more sophisticated role to the public, which the funnel model tends to reduce to nothing more than a passive source of external support. A theory or a scientific finding may consequently enjoy different status and robustness at different levels of communication. Thus the Big Bang may represent *the* explanation of the origin of the universe in the popular domain despite the doubts and distinctions expressed in the specialist one.

Interesting in this regard is the ambivalence of scientists towards situations characterized by deviation and, in general, towards their relations with the public. While deviation may be an opportunity to evade the rules and constraints of the popularization process, it is often regarded with suspicion by the specialist community. When scientific problems are pushed into the public arena, they lose some of the special status that they may still enjoy in such popularization frames as the scientific journals or the science sections of newspapers. They may, for example, be subject to problem concatenation processes or undergo 'life cycles' like all other issues of public interest: scientific theories, indeed, may in the end be likened to political doctrines and value judgements. Moreover, they can presumably also be manipulated and introduced into the public arena by actors external to the scientific community, like journalists, policy-makers or the leaders of movements and associations.

This helps explain the growing efforts by scientists to extend their control over communication with the public. Scientific institutions organize seminars on these matters and invite journalists to 'live laboratory life' for brief periods so that the standards of science communication are improved;[20] researchers write booklets advising their colleagues on how to handle the media.[21] Research institutes now make much use of public relations offices and similar devices, not to exclude the possibility of deviations (which would be difficult to achieve) but to extend the scientific community's control over recognition of 'crises' and over the activation of deviation processes so that the latter can be put to ad hoc use or, instead, criticized. To recall the 'double game' which Latour takes to be as distinctive of modernity – mixing science and society in practice but keeping them separate in theory – one notes that scientists often engage in deviation (i.e. public communication as part of the process by which a scientific fact is produced) but camouflage it as popularization (i.e. the diffusion of scientific knowledge with pedagogic intent) (see Chapter 6). Many of the misunderstandings that surround the debate on the public communication of science probably arise because popularization expectations are attributed to communications which, in reality, perform deviation functions – i.e. they serve to regulate the scientific debate for 'internal' purposes – and vice versa.

To draw an analogy with another theme treated in previous chapters, deviation with respect to popularization can be considered *à la* Kuhn equivalent to a scientific revolution with respect to normal science (see Chapter 2). There exists, in fact, a tension within the scientific community between the institutionalization of deviation – i.e. its absorption into ordinary expository practice (popularization) in order to prevent its 'uncontrolled abuse' – and its defence as a sort of 'emergency exit' for certain situations, and as a potential source of scientific change and innovation.

Notes

1. Several initiatives have been mounted in different countries to promote scientific knowledge, allocating funds to projects in this area and promoting activities such as 'science weeks' and 'science festivals'. Since 1999 the European Commission has launched specific funding schemes within the Framework Programme to encourage 'public awareness of science and technology'.
2. Cited in Pais (1982: 309).
3. See for example Friedman et al. (1986), Bettetini and Grasso (1988).
4. On the mad cow disease case see Kitzinger and Reilly (1997), Jasanoff (1997), Bucchi (1999); the debate on cloning in the Italian daily press has been studied by Neresini (2000).
5. Cf. Lewenstein (1995), Bucchi and Mazzolini (2003). Casadei (1991), for example, has conducted comparative lexical analysis of popular science texts, manuals and specialist articles on physics, finding entirely similar levels of technicality in the three genres, with the maximum level not in the specialist texts but in the manuals.
6. Cf. Hansen (1992), Peters (1995).
7. One of the most famous studies in the area, that conducted in 1991 by the National Science Foundation in the US, concluded, for example, that more than 90 per cent of the American and English populations could be considered as scientifically illiterate.
8. Cf. Gaskell and Bauer (2001).
9. Cf. Wynne (1995).
10. Cf. Gaskell et al. (2000), Gaskell and Bauer (2001), Bucchi and Neresini (2002).
11. More recently, interesting work has begun to appear in the area of science representation in fiction (Kirby, 2003; Massarani, 2002)
12. Similar conclusions are reached in a study on US scientists by Dunwoody and Scott (1982).
13. Cf. Cloître and Shinn (1985), Hilgartner (1990).
14. That is, the test detected the disease in only 15–20 per cent of subjects suffering from full-blown syphilis.
15. For this approach applied to the analysis of politics in the mass media see, for instance, Rositi (1982).
16. Macdonald and Silverstone (1992).
17. Collins describes these situations as 'distortions of the core set' (Collins, 1988).
18. The acronym initially used by researchers, GRID (Gay Related Immunodeficiency Disease), was abandoned under pressure by American homosexual activists and replaced with the term AIDS. Cf. Grmek (1989), Epstein (1996).
19. Food and Drug Administration, the authority responsible for testing medical drugs before they can be marketed in the US.
20. See, for instance, the EICOS initiative designed to give 'hands-on' laboratory experience to European science journalists (www.eicos.mpg.de).
21. For example, the *New England Journal of Medicine* advises researchers as follows: 'If you feel trapped, obfuscate: it will get cut if it's too technical' (cited in Nelkin, 1994: 31).

8 A new science?

1 A changing science

On 14 March 2000, the British Prime Minister, Tony Blair, and the President of the United States, Bill Clinton, issued a joint statement in which they applauded 'the decision of scientists working on the Human Genome Project to release raw fundamental information about the human DNA sequence and its variants rapidly into the public domain'. The two leaders concluded by urging researchers around the world to adopt this policy of rapid publication (Danchin, 2000). The next day, shares in Celera Genomics Inc. – a private company so active in the field of human gene mapping as to become a serious competitor against the public consortium of research institutes – fell sharply together with those of numerous biotechnological companies, and with them the Nasdaq technological stock index. On 6 April 2000, Celera announced completion of the entire genomic sequence of a single male individual. The announcement received blanket coverage by the news media and Celera shares rose by 40 per cent in a few hours. The public consortium responded by announcing the detailed sequencing of three human chromosomes roughly corresponding to 11 per cent of the overall human genome.

Finally, on 26 June 2000 both the scientists running the public consortium and Craig Venter of Celera were invited to the White House, where they shook hands with President Clinton and promised him and the British Prime Minister Blair – connected by videoconference – that they would proceed with rapid and joint publication of the human genome map. Less than one year later, on 12 February 2001, the front page of the *New York Times* announced that the publicly funded Human Genome Project and the private company Celera had mapped the bulk of the human genome – a map consisting of about 30,000 genes, many fewer than the 100,000 expected.

126 A new science?

What is striking about this episode is not just the intervention in first person by two heads of state – an event of great rarity in the history of science, the only equivalent being the 1987 agreement between Chirac and Reagan to share the credit for (and profit from) the discovery of HIV equally between France and the US – but the fact that science, politics and business by now seem inextricably connected and able to influence each other reciprocally.

Our journey through the sociology of science has almost reached its end. We have seen that there is no single sociology of science, but rather a plurality of approaches, theoretical stances and empirical methods. What sociology, therefore, and what science? When Merton conducted the first social studies of science in the post-war years, science was indubitably very different from what it had been at the beginnings of the industrial revolution: namely the science of Galileo and Newton so often invoked as the symbol of science *tout court*, despite historical and disciplinary differences. However, we cannot ignore – to conduct an exercise in reflexivity certainly more modest than that urged by the 'strong programme' – that in the approximately five decades of sociology of science examined in previous chapters, the object of study – science – has undergone profound changes.

In this final chapter I shall exemplify my argument by making reference to the project to map the human genetic code, one of the most ambitious enterprises, and with the widest social implications, ever undertaken in the history of science. It is a project that lies at the core of significant transformation in the ways in which scientific knowledge is produced, distributed and utilized, and especially in the role played by biology in scientific research.

2 From the double helix to three billion steps

In 1953, when the 25-year-old American microbiologist James Watson and his English colleague Francis Crick announced the discovery of the 'double helix' structure of DNA, biology was a discipline practised individually, or at most in small groups, and often in ill-equipped laboratories. Watson and Crick carried out the research that led to their discovery in a tiny office at the Cavendish Laboratory in Cambridge; and some years after the discovery, Crick was still working in a bicycle shed.

The situation was very different with physics, which at that time absorbed most of the public funds allocated to basic research by the leading industrialized countries. And yet the profound changes today taking place in science – and which some consider to be a 'second scientific revolution' – began in precisely those years.

After the military victory over Japan, the American authorities immediately decided to develop a programme of scientific collaboration with the country that had only just previously been their enemy. One priority of the programme was to study the mutagenic effects of nuclear radiation – presumably in an attempt to heal the wounds caused by the bombing of Nagasaki and Hiroshima that had so profoundly marked Japan and shaken world public opinion. Genetics soon proved essential for the understanding of these mutations and, as a consequence, somewhat oddly, the US federal agency responsible for nuclear programmes, the Department of Energy (DoE), found itself playing a leading role in biological research as well.

In the course of the 1980s, the priorities of American research policy changed in concomitance with the shifting patterns of international politics. While on the one hand tension between the two blocs seemed to be subsiding after the years of the Cold War, on the other there arose a perception of Japan as a threat to American economic supremacy. It was also realized that competitiveness could only be restored to the American economy by concentrating on one of the strengths of its Japanese rival: scientific and technological development.

Robert Sinsheimer, a molecular biologist about to become Chancellor of the University of California at Santa Cruz, discovered that 36 million dollars had been allocated to the construction of an optical telescope at his university. Santa Cruz was at the same time being appraised as a possible site for an enormous particle accelerator, the Superconductor Supercollider, which would have cost several millions of dollars. Sinsheimer was immediately prompted to ask why this money should not be given to biology: 'It was thus evident to me that physicists and astronomers were not hesitant to ask for large sums of money to support programmes they believed to be essential to advance their science' (cited in Wilkie, 1993: 76).

Thus, in 1985, Sinsheimer organized a meeting with some of the leading molecular biologists in the US, the purpose of which was to examine the possibility of applying for funds to set up an institute for the study of the human genome at Santa Cruz. Some scientists immediately voiced their concerns over the exorbitant cost of the project, especially in view of the still inadequate state of knowledge on the human genome. Estimates made at the time envisaged at least 15 years of work and three billion dollars of expenditure, with the fear that the project would remove funds from other research projects already begun and now close to yielding concrete results. In other words, biologists were unaccustomed to the style of Big Science; they were afraid of losing their independence and, unlike the physicists, did not have a well-established tradition of international collaboration and the division of labour into small sub-projects which could also utilize students and postgraduates.

Further concerns were expressed about the social and political implications of the project. Information about a person's genetic make-up might lead to discrimination against him or her by employers and insurance companies, or indeed to the eugenic policies historically associated with the memory of Nazism.

The Department of Energy was quick to profit from the hesitations of the scientific community. Although funding for nuclear weapons research had not yet been cut back to its present levels, the then director of the DoE, Charles DeLisi, saw genetic research as justifying the existence of institutions like his well into the future, should the role and the political and public importance of nuclear research be retrenched. During a conference held at Santa Cruz, DeLisi pledged the DoE's large-scale commitment to the human DNA sequencing project. The DoE had well-equipped research facilities and could divert funds out of its large reserve for nuclear research – five and a half million dollars in 1987 – to the project. The calculating power of the DoE's computers, developed and tested for physics, would be crucial for genetic research, he said: in fact, the two main nuclear research centres, Los Alamos and Livermore, had already been working on a 'genetic library' for a number of years.

It was not long before the scientific community voiced its hostility. A number of particularly 'visible' scientists – among them the Nobel prize-winner David Baltimore and James Watson himself – were convinced of the project's importance but objected to its being controlled by the DoE bureaucracy. Genome research should, rather, be guided 'by the perceived needs of science' (Wilkie, 1993: 81). The solution was to develop the project under the aegis of the National Institutes of Heath (NIH). The intervention by Watson, now a scientific and public celebrity, was decisive, and in 1988 he was appointed director of genome research at the NIH and given a budget of almost 30 million dollars. Watson later explained his decision to accept the post as follows: 'I realized that only once would I have the opportunity to let my scientific life encompass the path from double helix to the 3 billion steps of the human genome' (cited in Wilkie, 1993: 83).

The next year found Watson in charge of a National Center for Genome Research and with an allocation of 60 million dollars; in

1991, the Center had a 30-strong staff and a budget of more than 100 million dollars. The development of similar projects in other countries, notably Britain and France, persuaded genome researchers that their efforts must be coordinated on an international scale. In 1988 a first association entitled HUGO (Human Genome Organization) was created for the purpose of allotting tasks among the various laboratories involved. It was followed in 1990 by the HGP (Human Genome Project) international consortium, which comprised 16 laboratories and had a budget of around three billion dollars, two billion of which were provided by the US government. However, this was not enough to dispel the doubts of some scientists: in 1991, a group headed by the members of the Department of Molecular Genetics at the Harvard Medical School wrote a letter to the journal *Science* in which they argued that there was no justifiable reason for devoting so much money to the genome project. They stressed that 90 per cent of the human genome was of no interest and accused the senator sponsoring the project, Pete Domenici of New Mexico, of only doing so to favour the Los Alamos laboratories situated in his state (Kevles and Hood, 1992).

3 From academia to the stock exchange

The announcement of the complete sequencing of the human genome by Craig Venter, president of Celera Genomics, provoked surprise and consternation. How could a private company like Celera compete with a public international consortium like HGP, which had been in operation for a decade and was financed with billions of dollars by some of the richest industrialized countries? And how could it be possible for information on our genes to be in the hands of a private organization?

The explanation for this phenomenon, which today causes such concern, resides in developments that have taken place during the course of the last two decades.

After a period during which regulation and control of scientific and technological research had been emphasized, in the 1980s economic development – albeit in new form – was once again the central issue of debate on research policies. This shift of emphasis was due in part to the return to power in both the US and Great Britain of conservative governments more concerned with business culture and the development of markets than financial intervention by the state, and in part to a new 'Sputnik effect', which this time consisted in the competitive threat raised by the countries of the Far East, especially

Japan. Western experts identified the success of the Japanese economy as stemming from such crucial factors as an ability to manage long-period plans and forecasts and, in particular, Japan's integration of research policy with industrial policy.

New provisions issued by the US administration allowed universities to register and commercially exploit patents resulting from research financed by public funds. In particular, a series of constraints were removed from applied research on recombinant DNA, which attracted conspicuous investments by numerous multinationals. Congress and institutions such as the National Science Foundation, moreover, encouraged joint research by industry and universities by offering financial incentives and tax relief.

In 1980, a sensation was caused when a Californian biotechnology company, founded only four years previously by a university professor of genetics, Herbert Boyer, launched a public share offer. In the following year, MIT received a donation of 125 million dollars from a businessman to host the Whitehead Institute for research in molecular biology. Fully 25 per cent of the patents granted to American universities between 1969 and 1991 were awarded in 1990 and 1991, while the period 1987–1991 saw a 100 per cent increase in patenting agreements between universities and industry. Cooperation agreements between federal laboratories and companies rose by 900 per cent in the same period, with the opening of more than 1,000 mixed university/industry research centres. The percentage of academic research financed by the federal government diminished from 68 per cent in 1980 to 56 per cent in 1993.

Craig Venter had initially worked on the public genome project at the National Institutes of Health. However, he was irked by the slowness of the methods selected for the sequencing, hindered as they were by the sluggishness of the public bureaucracy and squabbling among the organizations involved (the DoE and the NIH especially). He realized that the scale on which molecular biologists customarily worked must change: they would have to 'think big, in industrial terms', and that there was room for a private company. Venter invented a method which accelerated the process by concentrating on RNA, the molecule used by the cells to identify the parts of the DNA that manufacture proteins. He then devised what is known as the 'shotgunning' technique based on the random fragmentation of chromosomes, which are then decoded and reassembled. He obtained 70 million dollars from private investors and, together with his colleague William Haseltine, founded two organizations. One of them was industrial (Human Genome Sciences) while the other was non-profit-making, the TIGR (The Institute for Genome Research), since this would enable Venter to enter into contracts with public institutions like the DoE and the NIH. In 1995, one year after the foundation of the TIGR, Venter announced that he had completed the sequencing of the genome of the *Haemophilus influenzae* bacterium. He then unexpectedly left TIGR and, in 1998, set up a new society – Celera – with funding, and especially technology, provided by Perkin-Elmer, a scientific apparatus company, the market leader in sequencing machines and associated with the computer colossus Compaq. Venter later declared that Celera would eventually make the entire sequence of the human genome freely available on the Internet, after patenting 100-odd genes of especial importance for the development of drugs and selling them 'on a non-exclusive basis' to pharmaceutical companies.

4 From specialist papers to the front pages of the newspapers

In 1980, Michael Gottlieb, a young immunologist working at the University of California at Los Angeles Hospital, noticed a number of patients suffering from a distinctly odd syndrome: five cases under his observation, all homosexuals, besides displaying symptoms like diarrhoea and weight loss had also developed a rare form of pneumonia. Two of them died shortly afterwards. Gottlieb sensed that he was in the presence of an emergency. He contacted an editor at the *New England Journal of Medicine*, one of the most prestigious medical periodicals in the US, explained the gravity of the situation and asked how long the journal would take to publish a report warning of the emergency. The editor replied that it would require at least three months, during which time Gottlieb must not make any statement to the media (if he did, publication would be cancelled), and in any case there was no guarantee that his piece would be published. He advised Gottlieb to contact the Centers for Disease Control.[1] Gottlieb accordingly prepared a short account entitled 'Pneumocystis Pneumonia in Homosexual Subjects – Los Angeles' and sent it to the CDC for publication in their weekly bulletin on mortality and disease, the *Morbidity and Mortality Week Report*. When it arrived, there was much discussion at the CDC on the advisability of publishing the account, in particular because of its reference to the fact that the subjects affected were gay. In the meantime, further cases of the syndrome were reported. A correspondent claimed that numerous doctors in New York had come across similar cases but probably did not want to talk about them because they had articles forthcoming in medical journals. The gay daily *New York Native* had given a certain prominence to these cases, but the local health district had hastened to deny the story. In the end, however, Gottlieb's article was published on one of the bulletin's inside pages, with all references to homosexuality removed.

This minor episode, which marked the beginning of another scientific affair of great public concern, namely AIDS, highlights the three cardinal points around which the communication of science rotates: secrecy (to protect one's discoveries and prevent their plagiarism by other researchers), discussion of ideas among colleagues and the sharing of these ideas with the general public and their adaptation to the social and political context. A perhaps idealized conception of the scientific enterprise and its relationship with other social dynamics has long conceived these cardinal points as arranged in a sequence: the solitary and confidential development of theories and experiments, discussion with other specialists, and then dissemination of the results through the media and educational institutions.

However, the development (not only quantitative) of scientific activity and the profound changes that have taken place in its organization and the role of factors traditionally considered external to it, like pressure groups and the mass media, mean that intersection, tension and even conflict among the three points of the triangle grow increasingly frequent. As the Genome Project well demonstrates, simultaneously at work may be pressures to keep a discovery secret, to obtain the cooperation of competing scientists, and to publicize the discovery in order to inform the public or simply to gain visibility, legitimacy and, in the end, economic resources.

Added to this is the huge impact of the Genome Project on the public imagination. 'The book of life', 'map', 'cartography', 'the instruction manual for our species': the wealth of metaphors used to describe the undertaking highlight its public resonance. Perhaps more than any other scientific enterprise, the Project has incorporated this dimension from its beginnings. As we have seen, considerations of political and social expediency initially hampered its acceptance by scientists. Thereafter social scientists and philosophers were included in the Project to study its implications for ethics and society. On the other hand, those directly involved in the Genome Project have widely exploited its visibility. It has been also the association, emphasized mainly in the public domain, between genetics and the treatment of diseases that has ensured the Project's success and acceptance despite the opposition of numerous specialists.

With its repeated series of announcements of partial, imminent or simply promised achievements throughout its history, the human genome mapping enterprise fitted ideally with the media need for – and responsiveness to – specific events. The *New York Times*, for instance, featured 1,069 articles on the human genome between 1996 and 2001; somewhat surprisingly, the peak in coverage by the newspaper did not coincide with the final announcement, but with the 2000 political intervention at the highest level and the consequent promise by both research groups that the conclusion of the enterprise was approaching.

One crucial problem seems to be the control and transparency of the information collected by the Genome Project: hence Clinton and Blair's recommendation that the entire map should be released into the public domain. The public consortium has consequently accelerated its work schedule to move this release forward and compel Celera to do the same. But while secrecy and the 'private' character of knowledge raise obvious dilemmas, the unfiltered dissemination of scientific results does not seem entirely unproblematic either, especially in a sector like genetics. Mention has already been made of the risk that genetic information will be used for discriminatory purposes; and complex issues with regard to the handling of such information by those directly concerned have already arisen in the therapeutic field. In Great Britain, for example, the wide media coverage of the discovery of a gene believed able to cause breast cancer – even if in only 5 to 10 per cent of cases – prompted numerous women with cases of this type of tumour in their families to undergo prophylactic mastectomy.

In the course of the Project's history, the conflict between secrecy and transparency has repeatedly arisen at other levels as well. In 1988, for example, Watson caused consternation by proposing that part of the work on the genome should be allocated to the Soviet Union. In 1991, for two months during the Gulf War, the US Department of Trade blocked access to the NIH computers by foreign researchers collaborating on the Project, for fear that information might be used for biological warfare (Wilkie, 1993: 92).

And while it is true that Celera has based a good deal of its work on sequences already obtained by the public consortium – which naturally calls into question its right to patent the results – it should also be pointed out that the American public institutions have been the first to register patents in this field and are still their main beneficiaries.

5 A map of contemporary science?

The story of the Genome Project exemplifies a number of ongoing patterns of change in contemporary science. Created for political purposes, as was typical of research policy in the post-war period, the Project was subsequently developed in coincidence with profound changes in the organization of scientific research; a 'post-academic' science which, compared to the past, is now characterized by a greater proximity to the contexts of its application, by the marked intersection of disciplinary fields, by the heterogeneity of the actors and institutions involved, and by what commentators term 'reflexivity' and 'social accountability' (Gibbons et al., 1994; Ziman, 2001).

Hence, the presence of Craig Venter alongside politics and public and university research does not signal the growing importance of the industrial component alone, but all the above features together. Progress in the Genome Project would have been impossible had biological research not crossed paths with the information sciences. And this was not only from the point of view of the availability of technological equipment: the concept of 'information', now crucial for molecular biology originated in cybernetics. Physics, moreover, had provided the strategic and organizational model for molecular biology, which has applied its lessons by seeking to study complex organisms on the basis of their essential constitutive elements. Genome research has also demonstrated the extent to which research is increasingly embedded within heterogeneous social networks comprising politicians, businessmen, journalists and even young computer hackers – Napster peer-to-peer technology for sharing resources, for instance, was one of the models used by the human genome researchers to coordinate several sequencing centres (Merriden, 2001).

From the organizational standpoint, the Genome Project can be viewed as in some way equivalent to the Manhattan Project for physics, a 'quantum leap' from a discipline of middling importance to one which epitomizes the new 'big science': a 'big science' as a collective endeavour carried forward by huge and complex organizations far in excess of Price's predictions and which is set to redefine the nature itself of the scientific profession – the articles which reported the results of mapping the human genome in the journals *Nature* and *Science* were signed respectively by 275 and 250 authors – but, above all, a 'big science' which ousts the old 'military–industrial' complex and replaces it with a new 'academic–industrial–governmental' complex (Etzkowitz and Webster, 1995) in which venture capitalists set the research agenda, researchers act as entrepreneurs, and politicians confusedly mediate between the two sides.

The relationship between researchers and industry is undoubtedly one of the most striking trends of recent years, and it significantly characterizes the current organization and policy of research. In areas like microelectronics, nanotechnology and biotechnology, especially, one witnesses an unprecedented interweaving between research and the market whereby 'scientific knowledge is transformed into economic activity' (Etzkowitz and Webster, 1995). More and more universities are opening offices to patent their research results, and equally frequent are contracts by which companies and specific sectors utilize university services and resources. And then to be noted is the recent phenomenon of the creation by universities and research institutes of for-profit, spin-off enterprises to exploit the proceeds from their research projects. Of modest proportions in the past, the funding of public research by industry now accounts for between 10 and 15 per cent of the entire public research budget in countries like Germany and Sweden. The total R&D expenditure in Finland rose by around 1 per cent between 1994 and 2001, an increase almost entirely due to private research investments in high technology sectors (OECD, 2002). The University of Harvard predicts that by 2010 more than one-quarter of its economic resources will be provided by industry (Etzkowitz and Webster, 1995). It is calculated that around 64 per cent of research world-wide is financed by companies and that almost 70 per cent of it is performed by the companies themselves, albeit with wide differences among countries – ranging from Italy, where state financing is still predominant, to Japan, where only 18.7 per cent of research funds derive from the public sector and 73.4 per cent from industry.

This 'second academic revolution' (Etzkowitz, 1990), whereby intellectual property is considered to be private property, is redefining the role of the scientist and eliminating the traditional 'division of labour' that recompensed the researcher with reputation and industry with profits. 'I can do good science and make money' was the telling summary of this transition offered by a molecular biologist interviewed by Etzkowitz (Etzkowitz and Webster, 1995: 489). The idealized ethical code of the professional scientist (Merton, 1942) based on the principles of communism and disinterestedness has given way to 'a new normative structure of science . . . reflecting the transformation of science from a relatively minor institution encapsulated from social influence to a major institution that influences and is influenced by other social spheres' (Etzkowitz and Webster, 1995: 488).

Similar processes redefine not only the relationship between the scientific and economic spheres but the features themselves of the scientific enterprise and the social role of the scientist. Thus,

136 *A new science?*

while industrial consultancy work by a university professor was once regarded as 'extraneous' to his/her role, it may in this new framework be an integral part of his/her research and one of his/her institutional duties. Moreover, peer review of work by colleagues may clash with the need for secrecy imposed by the possible commercial exploitation of research, which increases the centrifugal pressures toward private collaboration and financing.

'To whom does scientific knowledge belong?' becomes the increasingly crucial question at the intersection between science and society. Post-academic science challenges both the traditional norm of *communalism* – a pillar of modern science according to which 'research results do not count as scientific unless they are reported, disseminated, shared and eventually transformed into communal property' (Ziman, 2001: 110) – and the representation of technological invention as an individual enterprise on which the inventor should retain copyright that has become commonplace in our societies since the late nineteenth century.[2]

Interestingly, pressures for the privatization, patenting and commercial exploitation of research results are counter-balanced by initiatives aimed at freeing such results from the constraints characteristic of academic science, like printed publication. The exponential increase in the number of refereed journals – a total of around 20,000 is currently estimated, with more than two million articles published every year – and their cost and, moreover, their concentration in the hands of a few academic publishers, has made it impossible for even the richest academic libraries to purchase and archive them all.[3]

In 1994 Steven Harnad, a psychologist at the University of Southampton, sent an email to an electronic discussion list in which

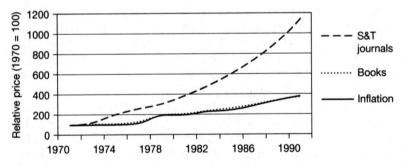

Figure 8.1 Price growth of scientific journals, compared with books and inflation
Source: Butler (1999)

he proposed that researchers worldwide should make their publications accessible through the internet, while complying with the academic standards guaranteed by peer-review. In 1991, even before Harnad's proposal, Paul Ginsparg, a physicist working at the Los Alamos national laboratories had set up *Arvix*, a database of High Energy Physics preprints and articles already accepted by journals but not yet published which today contains around 215,000 papers. Similar databases have also been set up for the biomedical sciences (*PubMed Central*), the cognitive sciences (*Cogprints*) and economics (*Repec*). In 2001, a group of American biologists launched the Public Library of Science campaign for the creation of a single integrated database of science articles freely accessible to both scientists and non-specialists. The campaign received support from over 30,000 scientists around the world. The importance of initiatives of this kind should not be viewed in economic terms alone, for they challenge the very concept of the scientific publication: if the basic recognition and reward unit of academic science has long been the printed article, what counts as a 'publication' in post-academic science? A paper archived in an online, refereed journal? A preprint circulated via the web? An informal communication to an electronic discussion list? Furthermore, the redefinition of science communication must take account of the ever-expanding and pervasive role of the mass media. It is possible to gather from the human genome story a lesson which also regards the 'mediatization' of science, where the research agenda is, at least in part, tailored to media needs so that the sequence of a specific gene or another partial achievement is announced by press conference almost every day.

More generally, the relatively low cost of the technology needed to conduct research and circulate its results – particularly in certain fields – makes it increasingly difficult to subject research activity to the traditional forms of control – both the internal ones carried out by the scientific community itself and sanctioned in terms of reputation and academic career, and those traditionally delegated to the political and justice spheres. Hence derive recent phenomena like 'biohacking': research on gene manipulation carried out somewhat amateurishly, not infrequently in informal contexts, outside universities and private research laboratories, with results being exchanged among informal web communities in the name of 'open source biocomputer science' (Eudes, 2002).

Indeed, one could speculate whether the expression 'scientific community' – which unquestionably suited the context of traditional, academic science – is still appropriate for networks of actors and practices not only increasingly porous to other social domains but also less internally consistent in terms of shared values, norms and routine practices.[4]

These transformations are connected with, and invariably accompanied by, profound changes in research, and also in the broader scenarios wherein such research is conducted. Other traditional distinctions now gradually losing their importance are those between science and technology, and between basic research and applied research, with the consequent development of increasingly intricate, but no less efficacious, patterns of interaction between scientific research and technological innovation (Faulkner, 1994). The distinctions among scientific sectors also grow increasingly blurred: in emerging areas like information and communication technology or biotechnology, indeed, research is by now entirely interdisciplinary. At the level of economic and productive macro-processes, the post-Fordist reorganization of capitalist production, with its emphasis on flexibility and outsourcing, assigns a more significant role to consultancy and research. As for research policy, one witnesses a widespread scaling-down of state support and more generally of the so-called 'project-grant system' – which awards research funding on the basis of specific projects and renews it according to the results and publications produced[5] – partially compensated by the right granted since the 1980s to American universities to take out patents on the results of their research. Scientific communities and academic institutions are increasingly involved in these local development projects, which have led *inter alia* to the creation of science and technology parks to promote and diffuse innovation. In this context, the public institutions do not lose their role but are redefined as the 'moderators/coordinators/facilitators' of relations among diverse subjects.

Another significant trend is the increasingly international dimension of scientific research, fostered by the spread of communications systems and testified to by the number of scientific articles published by international research groups – at present 25 per cent of the total – and of patents arising from inter-country agreements. The European countries in particular, driven by their need to counterbalance the scientific and technological hegemony of the US, have created a formidable array of international organizations like CERN (instituted in 1953 in Geneva), the European Space Agency (founded in 1975, with fourteen countries at present affiliated and a budget of three billion euros), the Joint European Tokamak (JET, founded in 1978) research centre for plasma physics, and other sectoral organizations like the European Molecular Biology Organisation (EMBO), the

European Molecular Biology Laboratory (EMBL), the European Southern Observatory (ESO), and the European Synchotron Radiation Facility (ESRF). Europe-level collaboration has also given rise to programmes for cooperation both scientific (the Framework Programme launched in 1984 and now in its fifth edition) and technological (the EUREKA programme started in 1985 in emulation of similar American and Japanese schemes to develop products, systems and services in all fields of advanced technology), and to the signing of bilateral research agreements between the European Union and various countries, including the US, China and Japan. At present, around 13 per cent of the budget allocated to research by the European Union member-states passes through Europe-level cooperation, financing the Framework Programme and bodies like CERN, EUREKA and ESA.[6] However, the impact of this 'globalization' of scientific research assumes different features in different countries. Contrasting to a country like Ireland, which seems especially able to profit from these processes in terms of international partnerships, is the relative 'isolation' of Japan, where no more than 1 per cent of firms are involved in industrial research agreements, and where the percentage of research funds of foreign origin accounts for only 0.1 per cent of the total.

Finally, the new 'big science' differs from its predecessor in that it must constantly negotiate its importance and social acceptability with a wide variety of publics and contexts. Paradoxically, in fact, the increasing economic importance of science has induced society to demand a legitimation of science which goes beyond its capacity to produce development and material affluence (Nowotny and Taschwer, 1996). Various forms of public mobilization have arisen in relation to specific scientific and technological initiatives, and, in general, there is increasing concern over the unforeseen and unwanted effects of scientific and technological progress on the environment and human health. The participation of citizen groups in decision-making with regard to these matters has been progressively recognized and institutionalized, especially in some countries of northern Europe and in the United States.

The attribution to science of greater 'social accountability', the use of scientific expertise to control research and development activities, and the creation of citizen panels where non-experts sit side by side with scientists and policy-makers, are among the most visible results of this process of increasing public attention and participation (Epstein, 1995; Bucchi, 1998b).

Indeed, it has been argued that common transformation processes – such as the increase in complexity and uncertainty or the emergence of a new economic rationality – have invested both science and society at large, one of the consequences being the increasing difficulty in establishing a clear demarcation between the two (Nowotny et al., 2001).

'This is you', the phrase with which the biologist Walter Gilbert, waving a CD-Rom, is wont to introduce his public lectures on the sequencing of DNA, encapsulates the meaning of a science that constantly photographs us, interprets us, interrogates us, patents us, and puts us to the test. More than a transformed science, according to some scholars this new scenario represents a 'New World Order Inc.' where biotechnology is now the promised land, and in which knowledge and private property, research and industry, intersect. The OncoMouse™, the first patented animal in the world,[7] metonymically represents technoscience as a whole and a new, colossal, scientific revolution in which transgenic animals perform the same role as transuranic elements during the Cold War (Haraway, 1997).

Faced with this scenario – indeed, *embedded* in this scenario – are those who argue that the tools of traditional sociology, and then those of the strong programme as well as laboratory studies, have been blunted. The proposal by the biologist and feminist anthropologist Donna Haraway is to replace theories about science with a plurality of 'positions' and 'situated knowledges'. The implosion of identities among economics, computer science and biology, and the renewal at once material and semiotic of the organisms brought forth by the New World Order Inc. blur the boundary between the technical and the political which constituted one of the central narratives of the scientific revolution and progress. Not only are guinea pigs, clones and cyborgs, the inhabitants of 'non-nature nature' like the Oncomouse™, those excluded and discarded by science, the subjects of this interstitial and situated knowledge; we are all, whether human or non-human, involved as 'non-innocent' authors of the new technoscience. One can no longer stand aloof from this technoscience like the modest witnesses to Boyle's experiments with his air pump – the genteel ladies not allowed to watch lest they try to save the birds suffocated during the experiments. We must 'squirm, organize, reveal, decry, preach, teach, deny, equivocate, analyze, resist, collaborate, contribute, denounce, expand, placate, withold'. The only thing we cannot do, Haraway concludes, 'in response to the meanings and practices that claim us body and soul is to remain neutral' (Haraway, 1997: 51).

Notes

1. A federal agency created in 1946 to combat infectious diseases which subsequently extended its purview to include non-infectious pathologies.
2. MacLeod (1996) documents how the definition of technological invention as result of individual genius rather than of a deterministic development – predominant until the mid-nineteenth century – is, in itself, the result of historical processes and social debates.
3. The US Association of Research Libraries (ARL) estimates that the average cost of library acquisitions grew by 8.8 per cent in each year between 1986 and 1998, for a total cost growth of 124 per cent (Case, 2001).
4. On the sociological concept of community, see the classic Tönnies (1877); on the scientific community, see Hagstrom (1965).
5. This system was first developed in the US and was widely used in that country during the years from the 1960s to the 1980s. Among its drawbacks were a tendency to exacerbate competition, the wasting of too much time on administrative procedures, its scant incentive to young researchers to develop broad-based research projects, and its tendency to distribute funds too narrowly (at the end of the 1960s, 25 per cent of grants were allocated to only ten universities). In order to remedy these shortcomings, other forms of financing were tried, either as alternatives to the system or in combination with it: the granting of institutional funds to universities which then distributed them internally according to their own criteria or – and this is the case of the National Institutes of Health, for example – *ad personam*, i.e. to researchers deemed to be particularly capable and reliable.
6. On the growth of European-level collaboration in the research field see Guzzetti (1995).
7. The product of genetic research financed by the multinational Du Pont at the Harvard Medical School, OncoMouse™ is a genetically modified mouse marketed in the catalogue of the Charles River laboratories in five different versions as carrying oncogenes able to mimic human breast cancers.

Suggested further reading and interesting websites

General

Jasanoff *et al.* (1995; the 2002 paperback edition has an updated bibliography) is the standard handbook in the field of Science and Technology Studies. It contains chapters on different STS areas and issues, each one written by a specialist of the specific theme. Nowotny and Taschwer (1996) is a two-volume reader that spans across the whole history of the field, from the sociology of knowledge classics up to the early 1990s; Biagioli (1999) is another reader, more focused on recent trends in the field.

Social Studies of Science – formerly *Science Studies* – and *Science Technology & Human Values* are the two leading journals in the field.

The website of the Society for Social Studies of Science is http://www/lsu.edu/ssss/.

1 The development of modern science and the birth of the sociology of science

Merton (1973) is a collection of Merton's most important papers on science. *Science, Technology and Society in Seventeenth-Century England* has been recently reprinted with a new introduction. Price (1963) is the classic reference for studies on the development of science; for an overview of such development see also Barnes (1985).

Updated information and data on R&D investments and R&D policy can be found at the website of the Organization for Cooperation and Economic Development (OECD), www.oecd.org and at the EU commission website, www.cordis.lu/rtd2002/indicators. News and documents on Science in Developing countries are available at www.scidev.net.

2 Paradigms and styles of thought: a 'social window' on science?

Kuhn (1962, 2nd edn 1969) and Fleck (1935, English trans. 1979) are highly recommended – albeit in some parts challenging – readings for those who wish to further explore the issues of change in scientific thought, paradigms and thought collectives.

3 Is mathematics socially shaped? The 'strong programme'

Shapin (1982) offers a wide-ranging overview of the early SSK studies. Bloor (1976, 2nd edn 1991) provides the standard presentation of the 'strong programme' (the new edition contains a response to the main critiques formulated against the programme). Barnes and Shapin (1979) is a collection of case studies from the history of science.

4 Inside the laboratory

Collins and Pinch (1993) is an accessible collection of case studies. Latour and Woolgar (1979) and Knorr Cetina (1981) are among the pioneering works in laboratory studies; Latour (1987) offers a detailed presentation of the actor-network approach. Pickering (1992) documents some of the theoretical and methodological debates characterizing the field.

5 Tearing bicycles and missiles apart: the sociology of technology

Collins and Pinch (1998) is an accessible collection of case studies in the area of technology. Bijker (1995) offers an ample and documented introduction to the Social Construction of Technology (SCOT) approach. The new edition of MacKenzie and Wajcman (1999) contains a number of interesting essays about different issues raised by the sociological analysis of technology as well as several case studies.

6 'Science wars'

Koertge (1998) is another collection of essays criticizing sociological studies of science; for further documentation on the Sokal Affair see the website http://physics.nyu.edu/faculty/sokal/.

7 Communicating science

Lewenstein (1995) and Gregory and Miller (1998) can be recommended for an overview of Public Communication of Science Studies.

Public Understanding of Science and *Science Communication* are the leading journals in the field.

The PCST (Public Communication of Science and Technology) network website, www.pcstnetwork.org, provides information on international conferences and events. In the area of public perception of science, a good example with regard to the specific issue of attitudes and opinions about biotechnology in Europe can be found at the website www.lse.ac.uk/lses.

8 A new science?

Gibbons *et al.* (1994), Haraway (1997), Nowotny *et al.* (2001), Ziman (2001) are all reflections – yet quite different from one another – on the recent transformations of science and its changing role within society.

References

Allison, P.D. and Stewart, J.A. (1974) 'Productivity Differences Among Scientists: Evidence for Accumulative Advantage', *American Sociological Review*, 39: 596–606.
Amsterdamska, O. (1990) 'Surely You're Joking, Monsieur Latour!', *Science Technology & Human Values*, 15: 495–504.
Ashmore, M. (1993) 'The Theatre of the Blind: Starring a Prometean Prankster, a Phoney Phenomenon, a Prism, a Pocket and a Piece of Wood', *Social Studies of Science*, 23, 1: 67–106.
Balmer, B. (1990) 'Scientism, science and scientists'. Unpublished research paper, Science Policy Research Unit, University of Sussex.
Barnes, B. (1974) *Scientific Knowledge and Sociological Theory*, London, Routledge & Kegan Paul.
—— (1982a) 'The Conventional Component in Knowledge and Cognition', in N. Stehr and V. Meja (eds) *The Knowledge Society*, Dordrecht, Reidel: 185–208.
—— (1982b) *T.S. Kuhn and the Social Sciences*, London, Macmillan.
—— (1983) 'Social Life as Boostrapped Induction', *Sociology*, 17, 4: 524–545.
—— (1985) *About Science*, Oxford, Blackwell.
—— (1990) 'Sociological Theories of Scientific Knowledge', in R.C. Olby et al. (eds), *Companion to the History of Modern Science*, London, Routledge: 60–73.
—— and Bloor, D. (1982) 'Relativism, Rationalism and the Sociology of Knowledge', in M. Hollis and S. Lukes (eds) *Rationality and Relativism*, Oxford, Blackwell.
—— and —— (1990) 'Relativismo, razionalità e la sociologia della conoscenza', in F. Dei and A. Simonicca, *Ragione e forme di vita. Razionalità e relativismo in antropologia*, Milano, Franco Angeli: 213–239.
—— and Dolby, R.G.A (1970) 'The Scientific Ethos: A Deviant Viewpoint', *Archives of European Sociology*, 11: 3–25.
—— and MacKenzie, D. (1979) 'On the Role of Interests in Scientific Change', in R. Wallis (ed.) *On the Margins of Science: The Social Construction of Rejected Knowledge*, 27, Keele: Sociological Review Monograph: 49–66.

References

—— and Shapin, S. (eds) (1979) *Natural Order. Historical Studies of Scientific Culture*, London, Sage.

——, Bloor, D. and Henry, J. (1996) *Scientific Knowledge: A Sociological Analysis*, London, Athlone.

Ben-David, J. (1971) *The Scientist's Role in Society. A Comparative Study*, Englewood Cliffs, Prentice Hall.

—— and Zloczower, A. (1962) 'Universities and Academic Systems in Modern Societies', *European Journal of Sociology*, 3, 1: 45–84.

Bernal, J. (1939) *The Social Function of Science*, London, Routledge.

Berridge, V. (1992) 'Aids, the Media and Health Policy', in P. Aggleton, P. Davies and G. Hart (eds) *AIDS: Right, Risk and Reason*, London, Falmer Press.

Bettetini, G. and Grasso, A. (1988) *Lo specchio sporco della televisione*, Turin, Fondazione Agnelli.

Biagioli, M. (ed.) (1999) *The Science Studies Reader*, New York, Routledge.

Bijker, W. (1995) *Of Bicycles, Bakelites and Bulbs*, Cambridge, MIT Press.

——, Hughes, T. and Pinch, T. (eds) (1987) *The Social Construction of Technological Systems. New Directions in the Sociology and History of Technology*, Cambridge, MIT Press.

Bloor, D. (1976) *Knowledge and Social Imagery*, London, Routledge & Kegan Paul, 2nd edn, Chicago, Chicago University Press, 1991.

—— (1982) 'Polyhedra and the Abominations of Leviticus: Cognitive Styles in Mathematics', in M. Douglas (ed.) *Essays in the Sociology of Perception*, London, Routledge (originally published in the *British Journal for the History of Science*, 11: 243–272, 1978).

—— (1983) *Wittgenstein: A Social Theory of Knowledge*, New York, Columbia University Press.

—— (1995) 'Idealism and the Social Character of Meaning', paper given at the University of California, Berkeley, 14 September.

—— (1996) 'Idealism and the Sociology of Knowledge', *Social Studies of Science*, 26, 6: 839–856.

—— (1999) 'Anti-Latour', *Studies in the History and Philosophy of Science*, 30, 1: 81–112.

Brown, J. (1989) *The Rational and the Social*, London, Routledge.

Bucchi, M. (1996) 'La scienza e i mass media: la "fusione fredda" nei quotidiani italiani', *Nuncius*, 2: 581–611.

—— (1997) 'The Public Science of Louis Pasteur: The Experiment on Anthrax Vaccine in the Popular Press of the Time', *History and Philosophy of the Life Sciences*, 19: 181–209.

—— (1998a) *Science and the Media. Alternative Routes in Scientific Communication*, London and New York, Routledge.

—— (1998b) 'La provetta trasparente: a proposito del caso Di Bella', *Il Mulino*, 1: 90–99.

—— (1999) *Vino, alghe e mucche pazze: la rappresentazione televisiva delle situazioni di rischio*, Rome, Eri/Rai.

—— (2001) 'Ricerca, politica della', in *Enciclopedia delle Scienze Sociali, Aggiornamento*, Rome, Istituto della Enciclopedia Italiana, vol. IX, 245–258.

—— and Mazzolini, R.G. (2003) 'Big Science, Little News: Science Coverage in the Italian Daily Press, 1946–1997', *Public Understanding of Science*, 12, 1: 7–24.

—— and Neresini, F. (2002) 'Biotech Remains Unloved by the More Informed', *Nature*, 416: 261.

Bulmer, R. (1967) 'Why is the Cassowary not a Bird?', *Man*, 2: 5–25.

Bunge, M. (1991) 'A Critical Examination of the New Sociology of Science', *Philosophy of the Social Sciences*, 21, 4: 524–560.

Butler, D. (1999) 'The Writing is on Web for Science Journals in Print', *Nature*, 397, 21 January: 195–200.

Butterfield, H. (1958) *The Origins of Modern Science*, London, Bell & Sons.

Cadeddu, A. (1987) 'Pasteur et la vaccination contre le charbon: una analyse historique et critique', *History and Philosophy of the Life Sciences*, 9: 255–276.

—— (1991) *Dal mito alla storia. Biologia e medicina in Pasteur*, Milano, Franco Angeli.

Callon, M. (1986) 'Some Elements of a Sociology of Translation: Domestication of the Scallops and the Fishermen', in J. Law (ed.) *Power, Action and Belief: A New Sociology of Knowledge?* London, Routledge & Kegan Paul.

—— and Latour, B. (eds) (1990) *La science tel qu'elle se fait*, Paris, La Découverte.

—— and Law, J. (1982) 'On Interests and their Transformations: Enrolment and Counterenrolment', *Social Studies of Science*, 12: 615–625.

Callon, M. and Rabeharisoa, V. (1999) *Le Pouvoir des Malades*, Paris, Presses de L'Ecole Nationale des Mines de Paris.

Casadei, F. (1991) 'Il lessico nelle strategie di presentazione dell'informazione scientifica', presented at Colloquio sulle Strategie Linguistiche dell' Informazione Scientifica, Rome, 9–10 December.

Case, M. (2001) 'The Impact of Serial Costs on Library Collections', *ARL Newsletter*, 218: 9.

Ceruzzi, P. (1999) 'Inventing Personal Computing', in D. MacKenzie and J. Wajcman (eds) *The Social Shaping of Technology*, Buckingham, Open University Press: 64–86.

Chia, A. (1998) 'Seeing and Believing. The Variety of Scientists' Responses to Contrary Data', *Science Communication*, 19, 4: 366–391.

Clemens, E. (1986) 'Of Asteroids and Dinosaurs: The Role of the Press in Shaping the Scientific Debate', *Social Studies of Science*, 16: 421–456.

—— (1994) 'The Impact Hypothesis and Popular Science: Conditions and Consequences of Interdisciplinary Debate', in William Glen (ed.) *The Mass-Extinction Debates: How Science Works in a Crisis*, Stanford, Stanford University Press.

References

Cloître, M. and Shinn, T. (1985) 'Expository practice: social, cognitive and epistemological linkages', in T. Shinn and R. Whitley (eds) *Expository Science*, Dordrecht, Reidel: 31–60.

—— (1986) 'Enclavement et diffusion du savoir', *Social Science Information*, 25, 1: 161–187.

Cohen, I.B. (1985) *Revolution in Science*, Cambridge, Harvard University Press.

Coleman, J.S. (1992) *Foundations of Social Theory*, Harvard, Belknap.

Collins, H.M. (1974) 'The TEA Set: Tacit Knowledge and Scientific Networks', *Science Studies*, 4: 165–186.

—— (1975) 'The Seven Sexes. A Study in the Sociology of a Phenomenon, or the Replication of Experiments in Physics', in *Sociology*, 9: 205–224.

—— (1981) 'Stages in the Empirical Programme of Relativism', *Social Studies of Science*, 11, 1: 3–10.

—— (1983) 'An Empirical Relativist Programme in the Sociology of Scientific Knowledge', in K. Knorr Cetina and M. Mulkay (eds) *Science Observed*, London, Sage: 85–113.

—— (1985) *Changing Order*, Chicago, University of Chicago Press.

—— (1987) 'Certainty and the Public Understanding of Science: Science on Television', *Social Studies of Science*, 17: 689–713.

—— (1988) 'Public Experiments and Displays of Virtuosity: the Core-set Revisited', *Social Studies of Science*, 18: 725–748.

—— and Pinch, T. (1993) *The Golem: What Everyone Should Know about Science*, Cambridge, Cambridge University Press.

—— and —— (1998) *The Golem at Large: What Everyone Should Know about Technology*, Cambridge, Cambridge University Press.

Cooter, R. and Pumfrey, S. (1994) 'Science in popular culture', *History of Science*, 32, 3: 237–267.

Crane, D. (1967) 'The Gatekeepers of Science. Some Factors Affecting the Selection of Articles of Scientific Journals', *American Sociologist*, 2: 195–201.

—— (1972) *Invisible Colleges: Diffusion of Knowledge in Scientific Communities*, Chicago, University of Chicago Press.

Curtis, R. (1994) 'Narrative Form and Normative Force. Baconian Storytelling in Popular Science', in *Social Studies of Science*, 24: 419–461.

Danchin, A. (2000) 'La storia del genoma umano', *Internazionale*, 341, 30 June: 20–24.

Dean, J. (1979) 'Controversy Over Classification: A Case Study from the History of Botany', in B. Barnes and S. Shapin (eds) *Natural Order: Historical Studies in Scientific Culture*, London, Sage.

Douglas, M. (1966) *Purity and Danger*, London, Routledge.

—— (1970) *Natural Symbols: Explorations in Cosmology*, Harmondsworth, Penguin.

Dröscher, A. (1998) 'The History of the Golgi Apparatus in Neurons from its Discovery 1898 to Electron Microscopy', *Brain Research Bulletin 47*, 199–203.

References 151

Dubois, M. (1999) *Introduction à la sociologie des sciences*, Paris, Presses Universitaires de France.
Dunwoody, S. and Scott, B. (1982) 'Scientists as Mass Media Sources', *Journalism Quarterly*, 59, 1: 52–59.
Dyer, R. (1999) 'Making "White" People White', in MacKenzie, D. and Wajcman, J. (eds) *The Social Shaping of Technology*, Buckingham, Open University Press: 134–140.
Eltzinga, A. and Jamison, A. (1995) 'Changing Policy Agendas in Science and Technology', in S. Jasanoff *et al.* (eds) *Handbook of Science and Technology Studies*, Thousand Oaks, Sage.
Epstein, S. (1995) 'The Construction of Lay Expertise: AIDS Activism and the Forging of Credibility in the Reform of Clinical Trials', *Science, Technology and Human Values*, 20, 4: 408–437.
—— (1996) *Impure Science: AIDS, Activism and the Politics of Knowledge*, Berkeley, University of California Press.
Etzkowitz, H. (1990) 'The Second Academic Revolution. The Role of the Research University in Economic Development', in S. Cozzens, P. Healy, A. Rip and J. Ziman (eds), *The R&D System in Transition*, Dordrecht, Reidel.
—— and Webster, A. (1995) 'Science as Intellectual Property', in S. Jasanoff *et al.* (eds) *Handbook of Science and Technology Studies*, Thousand Oaks, Sage: 480–505.
Eudes, Y. (2002) 'Le pirates du génome', *Le Monde*, 18 September.
Faulkner, W. (1994) 'Conceptualizing Knowledge Used in Innovation: A Second Look at the Science-Technology Distinction and Industrial Innovation', *Science Technology and Human Values*, 19, 4: 425–458.
Felt U., Nowotny, H. and Taschwer, K. (1995) *Wissenschafts-forschung. Eine Einführung*, Frankfurt, Campus.
Feyerabend, P. (1975) *Against Method. Outline of an Anarchist Theory of Knowledge*, London, New Left Books.
—— (1996a) *Ambiguità e armonia*, Rome, Laterza.
—— (1996b) 'Contro l'ineffabilità culturale. Oggettivismo, relativismo e altre chimere', in G. De Finis and R. Scartezzini, *Universalità e Differenza*, Milan, Franco Angeli.
Fleck, L. (1935) *Entstehung und Entwicklung einer wissenschaftliche Tatsache*, (English trans. T. Trenn (1979) *Genesis and Development of a Scientific Fact*, Chicago, University of Chicago Press).
Forman, M. (1971) 'Weimar Culture, Causality and Quantum Theory 1918–1927', *Historical Studies in the Physical Sciences*, Philadelphia, University of Pennsylvania Press, vol. III: 1–115.
Fox Keller, E. (1995) *Refiguring Life. Metaphors of Twentieth-Century Biology*, New York, Columbia University Press, 1995.
Frankel, E. (1976) 'Corpuscular Optics and the Wave Theory of Light: The Science and Politics of a Revolution in Physics', *Social Studies of Science*, 6: 141–184.
Friedman, S.M., Dunwoody, S. and Rogers, C.L. (eds) (1986) *Scientists and Journalists. Reporting Science as News*, New York, Free Press.

References

Gale, G. (1972) 'Darwin and the Concept of a Struggle for Existence: A Study in the Extrascientific Origin of Ideas', *Isis*, 63: 321–344.

Garfinkel, H. (1967) *Studies in Ethnomethodology*, Englewood Cliffs, Prentice Hall.

——, Lynch, M. and Livingston, E. (1981) 'The Work of a Discovering Science Construed with Materials from the Optically Discovered Pulsar', *Philosophy of the Social Sciences*, 11: 131–158.

Gaskell, G. and Bauer, M. (eds) (2001) *Biotechnology 1996–2000. The Years of Controversy*, London, The Science Museum.

—— et al. (2000) 'Biotechnology in the European Public', *Nature Biotechnology*, 18, 9: 935–938.

Gibbons, M., Limoges, C., Nowotny, H. et al. (1994) *The New Production of Knowledge. The Dynamics of Science and Research in Contemporary Societies*, London: Sage.

Gieryn, T. and Figert, A. (1990) 'Ingredients for a Theory of Science in Society. O-rings, Ice Water, C-Clamp, Richard Feynman and the Press', in S.E. Cozzens and T.F. Gieryn (eds) *Theories of Science in Society*, Bloomington, Indiana University Press.

Giglioli, P.P. and Dal Lago, A. (1983) *Etnometodologia*, Bologna, Il Mulino.

Gilbert, N. and Mulkay, M. (1982) 'Accounting for Error in Science', *Sociology*, 16: 165–183.

—— (1984) *Opening Pandora's Box: A Sociological Analysis of Scientific Discourse*, Cambridge, Cambridge University Press.

Goodel, R. (1977) *The Visible Scientists*, Boston, Little Brown.

—— (1987) 'The Role of the Mass Media in Scientific Controversies', in H.T. Engelhardt Jr and A.L. Caplan, *Scientific Controversies*, Cambridge, Cambridge University Press.

Gregory, J. and Miller, S. (1998) *Science in Public. Communication, Culture, and Credibility*, London, Plenum.

Grmek, M.D. (1989) *Histoire du SIDA*, Paris, Payot.

Grundmann, R. and Cavaillé, J.P. (2000) 'Simplicity in Science and its Publics', *Science as Culture*, 9, 3: 353–389.

Guzzetti, L. (1995) *A Brief History of European Union Research Policy*, Luxembourg, European Commission.

Hacking I. (1992) 'The Self-Vindication of the Laboratory Sciences', in A. Pickering (ed.) *Science as Practice and Culture*, Chicago, The University of Chicago Press.

—— (1999) *The Social Construction of What?* Harvard University Press.

Hagstrom, W.O. (1965) *The Scientific Community*, New York, Basic Books.

—— (1982) 'Gift Giving as an Organizing Principle in Science', in B. Barnes and D. Edge (eds) *Science in Context*, Milton Keynes, Open University Press: 21–34.

Hall, A.R. (1983) *Revolution in Science 1500–1750*, London, Longman.

Hall, B. (1996) 'Lynn White's "Medieval Technology and Social Change" After Thirty Years', in R. Fox (ed.) *Technological Change. Methods and Themes in the History of Technology*, Reading, Harwood: 257–263.

Hansen, A. (1992) 'Journalistic Practices and Science Reporting in the British Press', *Public Understanding of Science*, 1, 3: 111–134.
Haraway, D. (1997) *Modest_Witness@Second Millennium. FemaleMan© Meets Oncomouse^{TM}*, New York, Routledge.
Henderson, P. and Kitzinger, J. (1999) *The Human Drama of Genetics: 'Hard' and 'Soft' Media Representations of Inherited Breast Cancer*, Glasgow Media Centre Working Paper 85, 1999.
Heritage, J. (1984) *Garfinkel and Ethnomethodology*, Oxford, Polity Press.
Hess, D.J. (1997) *Science Studies: An Advanced Introduction*, New York, New York University Press.
Hesse, M.J. (1966) *Models and Analogies in Science*, Notre Dame, Notre Dame University Press.
Hilgartner, S. (1990) 'The Dominant View of Popularization', *Social Studies of Science*, 20: 519–539.
—— (1997) 'The Sokal Affair in Context', *Science Technology and Human Values*, 22, 4: 506–522.
Hughes, T.P. (1999) 'Edison and Electric Light', in D. MacKenzie and J. Wajcman (eds) *The Social Shaping of Technology*, Buckingham, Open University Press.
Jacobi, D. (1987) *Textes et images de la vulgarisation scientifique*, Bern, Peter Lang.
Jasanoff, S. (1997) 'Civilization and Madness: the Great BSE Scare of 1996', *Public Understanding of Science*, 6: 221–232.
——, Markle, G.E., Petersen, J.C. and Pinch T. (eds) (1995) *Handbook of Science and Technology Studies*, Thousand Oaks, Sage.
Kevles, D.J. (1999) 'Pursuing the Unpopular: A History of Courage, Viruses and Cancer', in R.B. Silvers (ed.) *Hidden Histories of Science*, New York: The New York Review of Books.
—— and Hood, L. (eds) (1992) *The Code of Codes. Scientific and Social Issues in the Human Genome Project*, Cambridge, MA, Harvard University Press.
Kirby, D.A. (2003) 'Scientists on the Set: Science Consultants and the Communication of Science in Visual Fiction', *Public Understanding of Science*, 12, 3: 261–278.
Kitzinger, S. and Reilly, J. (1997) 'The Rise and Fall of Risk Reporting. Media Coverage of Human Genetic Research, "False Memory Syndrome" and "Mad Cow Disease"', *European Journal of Communication*, 3: 319–350.
Klaw, S. (1968) *The New Brahmins*, New York, Morrow.
Knorr Cetina, K. (1981) *The Manufacture of Knowledge: An Essay on the Constructivist and Contextual Nature of Science*, Oxford, Pergamon.
—— (1995) 'Laboratory Studies: The Cultural Approach to the Study of Science', in S. Jasanoff, G.E. Markle, J.C. Petersen, and T. Pinch (eds) *Handbook of Science and Technology Studies*, Thousand Oaks, Sage: 140–165.
Koertge, N. (1998) *A House Built on Sand: Exposing Postmodernist Myths about Science*, Oxford, Oxford University Press.

Kohler, R.E. Jr (1972) 'The Reception of Eduard Buchner's Discovery of Cell-Free Fermentation', *Journal of the History of Biology*, 5: 327–353.

Kuhn, T.S. (1962) *The Structure of Scientific Revolutions*, Chicago: Chicago University Press (2nd edn, 1969).

Lakatos, I. (1976) *Proofs and Refutations. The Logic of Mathematical Discovery*, Cambridge, Cambridge University Press.

Latour, B. (1983) 'Give me a laboratory and I will raise the world', in K. Knorr-Cetina and M. Mulkay (eds) *Science Observed*, London, Sage.

—— (1984) *Les Microbes. Guerre et Paix*, Paris, Metailié.

—— (1987) *Science in Action*, Cambridge, MA, Harvard University Press.

—— (1991) *Nous n'avons jamais été modernes*, Paris, La Découverte (English trans. Latour, B. (1993) *We Have Never Been Modern*, London, Harvester Weatsheaf).

—— (1992) *Aramis ou l'amour des techniques*, Paris, Editions La Découverte.

—— (1995) *Le métier de chercheur regard d'un antropologue*, Paris, Inra.

—— and Callon, M. (1990) (eds) *La Science telle qu'elle se fait*, Paris, Editions La Découverte.

—— and Woolgar, S. (1979) *Laboratory Life. The Social Construction of Scientific Facts*, Princeton, Princeton University Press.

Layton, E.T. (1977) *Conditions of Technological Development*, in I. Spiegel-Rösing and D.J. de Solla Price (eds), *Science, Technology and Society*, London, Sage, 1977: 197–222.

—— (1988) 'Science as a Form of Action. The Role of the Engineering Sciences', in *Technology and Culture*, 28: 594–607.

Lewenstein, B. (1992a) 'Cold fusion and hot history', *Osiris*, second series, 7: 135–163.

—— (1992b) 'The Meaning of "Public Understanding of Science"' in the United States After World War II', *Public Understanding of Science* I: 45–68.

—— (1995) 'Science and the media', in S. Jasanoff et al. (eds) *Handbook of Science and Technology Studies*, Thousand Oaks, Sage: 343–359.

Liebowitz, S.J. and Margolis, S.E. (1995) 'Path Dependence, Lock-in and History', *Journal of Law, Economics and Organization*, 11, 1: 205–226.

Lupton, D. (1995) *The Imperative of Health*, London, Sage.

Macdonald, S. and Silverstone, R. (1992) 'Science on Display: The Representation of Scientific Controversy in Museum Exhibition', *Public Understanding of Science*, 1, 69: 87.

MacKenzie, D. (1976) 'Eugenics in Britain', *Social Studies of Science*, 6: 499–532.

—— (1978) 'Statistical Theory and Social Interests: A Case Study', *Social Studies of Science*, 8: 35–83.

—— (1993) 'Negotiating Arithmetic, Constructing Proof: The Sociology of Mathematics and Information Technology', *Social Studies of Science*, 23: 37–65.

—— (1996) 'How Do We Know the Properties of Artefacts? Applying the Sociology of Knowledge to Technology', in R. Fox (ed.) *Technological*

Change. Methods and Themes in the History of Technology, Reading, Harwood: 257–263.
—— (1999) 'Slaying the Kraken: The Sociohistory of a Mathematical Proof', *Social Studies of Science*, 29, 1: 7–60.
—— (2001) 'Physics and Finance: S-Terms and Modern Finance as a Topic for Science Studies', *Science, Technology & Human Values*, 26, 2: 115–144.
—— and Wajcman, J. (eds) (1999) *The Social Shaping of Technology*, Buckingham, Open University Press.
MacLeod, C. (1996) 'Concepts of Invention and the Patent Controversy in Victorian Britain', in R. Fox (ed.) *Technological Change. Methods and Themes in the History of Technology*, Reading, Harwood: 137–153.
Mannheim, K. (1925) 'The Problem of a Sociology of Knowledge from a Dynamic Standpoint', in H. Nowotny and K. Taschwer (eds) *The Sociology of the Sciences*, vol. I: 3–14.
March, J.P. and Olsen, J.P. (1989) *Rediscovering Institutions*, New York, Free Press.
Massarani, L. (2002) 'Review of O Clone', *Public Understanding of Science*, 11 April 2002: 207–208.
Mazzolini, R.G. (1988) *Politisch-Biologische Analogien in Frühwerk Rudolf Virchows*, Marburg, Basilisken.
Mazzotti, M. (1998) 'The Geometers of God. Mathematics and Reaction in the Kingdom of Naples', *Isis*, 89: 674–701.
Merriden, T. (2001) *Irresistible Forces. The Legacy of Napster and the Growth of the Underground Internet*, Oxford, Capstone.
Merton, R.K. (1938) *Science, Technology and Society in Seventeenth-Century England*, Bruges, St Catherine Press (fourth edn, with a new introduction, New York, Howard Fertig, 2001).
—— (1942) 'The Normative Structure of Science', repr. in *The Sociology of Science*, 1973.
—— (1952) 'The Neglect of the Sociology of Science', repr. in *The Sociology of Science*, 1973.
—— (1963) 'The Ambivalence of Scientists', repr. in *The Sociology of Science*, 1973.
—— (1968a) 'The Matthew Effect in Science', repr. in *The Sociology of Science*, 1973.
—— (1968b) *Social Theory and Social Structure* (original edn 1949), New York, Free Press.
—— (1973) *The Sociology of Science. Theoretical and Empirical Investigations*, Chicago, University of Chicago Press.
—— and Barber, E. (1963) 'Sociological Ambivalence', in E.A. Tiryakian (ed.) *Sociological Theory, Values and Sociocultural Change*, Glencoe, Free Press, 91–120.
Miller, S. (1994) 'Wrinkles, Ripples and Fireballs. Cosmology on the Front Page', *Public Understanding of Science*, 3: 445–453.

—— and Gregory, J. (1998) *Science in Public. Communication, culture, and credibility*, London: Plenum.

Mitroff, I. (1974) 'Norms and Counter-Norms in a Select Group of The Apollo Moon Scientists: A Case Study of the Ambivalence of Scientists', *American Sociological Review*, 39: 579–595.

Mulkay, M. (1974) 'Conceptual Displacement and Migration in Science: A Prefatory Paper', *Science Studies*, IV: 205–234.

—— (1979) *Science and the Sociology of Knowledge*, London, Allen & Unwin.

National Science Board (2000) *Science and Technology Indicators 2000*.

Nelkin, D. (1994) 'Promotional metaphors and their popular appeal', *Public Understanding of Science*, 3: 25–31.

—— and Lindee, S. (1995) *The DNA Mystique. The Gene as a Cultural Icon*, New York, Freeman, 1995.

Neresini, F. (2000) 'And Man Descended from the Sheep: The Public Debate on Cloning in the Italian Press', *Public Understanding of Science*, 9: 359–382.

Nowotny, H. and Taschwer, K. (eds) (1996) *The Sociology of Sciences*, 2 vols, Cheltenham, Elgar.

——, Scott, P. and Gibbons, M. (2001) *Re-Thinking Science. Knowledge and the Public in an Age of Uncertainty*, Cambridge, Polity Press.

OECD (1999) *Science, Technology and Industry Scoreboard 1999. Benchmarking Knowledge Based Economies*, Paris, OECD.

—— (2002) *Science, Technology and Industry Outlook*, Paris.

Ospovat, D. (1978) 'Perfect Adaptation and Teleological Explanation: Approaches to the Problem of the History of Life in the Mid-Nineteenth Century', *Studies in the History of Biology*, 2: 33–56.

Pais, A. (1982) *Subtle is the Lord. The Science and Life of Albert Einstein*, New York, Oxford University Press.

Perec, J. (1991) *Cantatrix Sopranica L. et autres écrits scientifiques*, Paris, Editions du Seuil.

Peters, H.P. (1994) 'Mass Media as an Information Channel and Public Arena', *Risk: Health, Safety & Environment*, 5, 241–250.

—— (1995) 'The Interaction of Journalists and Scientific Experts: Co-operation and Conflict Between Two Professional Cultures', *Media Culture & Society*, 17: 31–48.

—— (2000) 'Scientists as Public Experts', paper presented at the 6th PCST Conference, Geneva, 1 February.

Phillips, D.M. (1991) 'Importance of the lay press in the transmission of medical knowledge to the scientific community', *New England Journal of Medicine*, 11 October: 1180–1183.

Pickering, A. (1980) 'The Role of Interests in High-Energy Physics: The Choice between Charm and Color', in K. Knorr, R. Krohn and R. Whitley (eds) *The Social Process of Scientific Investigation*, Dordrecht, Reidel: 107–138.

—— (ed.) (1992) *Science as Practice and Culture*, Chicago: University of Chicago Press.

Pinch, T. (1999) 'Giving Birth to New Users: How the Minimoog was sold to Rock and Roll', paper presented at the Annual Meeting of the Society for the Social Studies of Science, San Diego, 28–29 October.

—— and Bijker, W. (1990) 'The Social Construction of Facts and Artefacts: Or How the Sociology of Science and the Sociology of Technology Might Benefit Each Other', in W. Bijker, T. Hughes and T. Pinch (eds) *The Social Construction of Technological Systems*, Cambridge, MA, MIT Press.

Pizzorno, A. (1986) 'Sul confronto intertemporale delle utilità', *Stato e Mercato* 16: 3–25.

Powell, W.W. and Di Maggio, P.J. (1991) *The New Institutionalism in Organizational Analysis*, Chicago, The University of Chicago Press.

Price, D.J. de Solla (1963) *Little Science, Big Science*, New York, Columbia University Press.

Raichvarg, D. and Jacques, J. (1991) *Savants et ignorants. Une histoire de la vulgarisation de sciences*, Paris, Editions du Seuil.

Raup, D. (1991) *Extinction. Bad Genes or Bad Luck?* Chicago: University of Chicago Press.

Rosen, P. (1993) 'The Social Construction of Mountain Bikes: Technology and Postmodernity in the Cycle Industry', *Social Studies of Science*, 23: 479–513.

Rosenberg, N. (1982) *Inside the Black Box: Technology and Economics*, Cambridge, Cambridge University Press.

Rositi, F. (1982) *I modi dell' argomentazione e l'opinione pubblica*, Turin, Eri.

Rossi, P. (1988a) 'Il fascino della magia e l'immagine della scienza', in P. Rossi (ed.) *Storia della Scienza*, Turin, UTET, vol. I: 31–57.

—— (1988b) 'Le istituzioni e le immagini della scienza', in P. Rossi (ed.) *Storia della Scienza*, Turin, UTET, vol. I: 3–29.

—— (1997) *La nascita della scienza moderna in Europa*, Bari, Laterza, 1997.

Rudwick, M.J. (1974) 'Poulett Scrope on the Volcanoes of Auvergne: Lyellian Time and Political Economy', *The British Journal for the History of Science*, 7: 205–242.

Seagall, A. and Roberts, L.W. (1980) 'A Comparative Analysis of Physician Estimates and Levels of Medical Knowledge Among Patients', *Sociology of Health and Illness*, 2: 317–334.

Shapin, S. (1979) 'The Politics of Observation: Cerebral Anatomy and Social Interests in the Edinburgh Phrenology Disputes', in R. Wallis (ed.) *On the Margins of Science: The Social Construction of Rejected Knowledge*, London, Routledge & Kegan Paul: 139–178.

—— (1982) 'History of Science and its Sociological Reconstructions', *History of Science*, 20: 157–211.

—— and Schaffer, S. (1985) *Leviathan and the Air Pump: Hobbes, Boyle and the Experimental Life*, Princeton, Princeton University Press.

Snow, C.P. (1960) *Science and Government*, Cambridge, Harvard University Press.

Sobel, D. (1995) *Longitude*, New York, Walker Books.

References

Sokal, A. (1996a) 'Transgressing the boundaries: Toward a transformative hermeneutics of quantum gravity', *Social Text*, 14, 1–2: 217–252.

—— (1996b) 'A Physicist Experiments with Cultural Studies', *Lingua Franca*, May/June: 62–64.

—— and Bricmont, J. (1997) *Impostures intellectuelles*, Paris, Editions Odile Jacob (English trans. Sokal, A. and Brimont, (1998) *Fashionable Nonsense: Postmodern Intellectuals' Abuse of Science*, New York, Picador).

Tönnies, F. (1887) *Gemeinschaft und Gesellschaft*, Leipzig Fues's Verlag (English trans. Loomis, C.P. (1957) *Community and Society*, New York: Harper & Row).

Weber, M. (1905) *Die protestantische Ethik und der Geist des Kapitalismus*, Tübingen, Mohr.

Weinberg, S. (1996) 'Sokal's Hoax', *The New York Review of Books*, 43, 13, 8 August: 11–15.

White, L. Jr (1962) *Medieval Technology and Social Change*, Oxford, Oxford University Press.

Whitley, R. (1985) 'Knowledge Producers and Knowledge Acquirers', in T. Shinn and R. Whitley (eds) *Expository Science: forms and functions of Popularization*, Yearbook Sociology of the Sciences IX, Dordrecht, Reidel: 3–28.

Wilkie, T. (1993) *Perilous Knowledge. The Human Genome Project and its Implications*, London, Faber & Faber.

Woolgar, S. (1988) *Knowledge and Reflexivity: New Frontiers in the Sociology of Knowledge*, London, Sage.

Wynne, B. (1979) 'Physics and Psychics. Science, Symbolic Action and Social Control in Late Victorian England', in B. Barnes and S. Shapin (eds) *Natural Order. Historical Studies of Scientific Culture*, London, Sage: 167–189.

—— (1982) 'Natural Knowledge and Social Context: Cambridge Physicists and the Luminiferous Ether', in B. Barnes and D. Edge (eds) *Science in Context*, Milton Keynes, Open University Press: 212–231.

—— (1989) 'Sheepfarming after Chernobyl: A Case Study in Communicating Scientific Information', *Environment Magazine*, 31, 2: 10–39.

—— (1995) 'Public Understanding of Science', in S. Jasanoff *et al.* (eds) *Handbook of Technology Studies*, Thousand Oaks, Sage: 361–389.

Young, R.M. (1969) 'Malthus and the Evolutionists: The Common Context of Biological and Social Theory', *Past and Present*, 43: 109–145.

—— (1973) 'The Historiographic and Ideological Context of the Nineteenth-Century Debate on Man's Place in Nature', in M. Teich and R.M. Young (eds), *Changing Perspectives in the History of Science*, London: Heinemann 344–438.

Ziman, J. (2001) *Real Science. What It Is, and What It Means*, Cambridge, Cambridge University Press.

Ziporyn, T. (1988) *Disease in the Popular American Press: The Case of Diphtheria, Typhoid Fever, and Syphilis, 1870–1920*, New York, Greenwood.

Zuckerman, H. (1977) *Scientific Elite: Nobel Laureates in the United States*, New York, Free Press.

Index of names

Abel, N.H. 38
Adams, J. 41
Agassiz, J.L.R. 52
Alvarez, L. 70,118
Amsterdamska, O. 75
Appel, K. 101, 106
Arago, F. 37
Ashmore, M. 59

Balmer, B. 118
Baltimore, D. 43, 128
Barber, E. 18, 19, 21
Barnes, B. 13, 18, 23, 31, 36, 37, 38, 40, 42, 46, 48, 81, 89, 92, 98, 100, 102, 103, 105
Battelli, V. 34
Baudrillard, J. 93
Ben-David, J. 12, 23, 56
Bergson, H. 93, 94
Bernal, J. 15
Berridge, V. 39
Berzelius 50
Bettetini, G. 123
Bijker, W. 84, 86
Blair, T. 125, 133
Blondlot, R. 51, 56, 59, 74
Bloor, D. 3, 38, 42, 48–58, 62, 93, 99, 100, 101, 102–104, 105
Boyer, H. 130
Boyle, R. 76, 13, 58, 96, 140
Brahe, T. 6
Bricmont, J. 94, 95, 96
Brown, J. 57
Bucchi, M. 44, 65, 70, 113, 119, 123, 139
Buchner, E. 38

Bulmer, R. 34, 36
Bunge, M. 55
Bush, V. 79
Butterfield, H. 12

Cadeddu, A. 33, 65
Callon, M. 70, 102
Casadei, A. 123
Cauchy, A.L. 38, 53, 55
Cavaillè, J.P. 118
Ceruzzi, P., 82
Chia, A. 30, 40
Churchill, W. 14
Clemens, E. 70, 118
Clinton, B. 125, 133
Cloître, M. 114, 115, 118, 123
Cohen, I.B. 23, 32, 40
Coleman, J.S. 106
Collins, H.M. 66–69, 86, 88, 90, 99, 102, 104, 117, 123, 144
Cooter, R. 121
Crane, D. 21
Crick, F. 22, 28, 126
Curtis, R. 120

Dal Lago, A. 76
Danchin, A. 125
Darwin, C. 25, 26, 43, 44, 52
Dean, J. 25, 69
De Lisi, C. 128
Descartes, R. 21
Diderot, D. 21
Di Maggio, P.J. 106
Dolby, R.G.A. 18
Doyle, A.C. 117
Douglas, M. 37

Index of names

Dröscher, A. 43
Duesberg, P. 93
Dulbecco, R. 43
Durkheim, E. 2
Du Toit, A. 32, 52
Dyer, R. 88

Edge, D. 3, 41, 48
Edison, T. 81, 97
Eltzinga, A. 15
Epstein, W. 76, 95, 121, 123, 139
Etzkowitz, H. 134, 135
Euler, L. 53–55

Faulkner, W. 138
Feyerabend, P.K. 2, 32, 93, 94, 95, 106
Feynman, R. 102, 106
Figert, A. 106
Flaubert, G. 21
Flauti, V. 47
Fleck, L. 2, 39, 40, 56, 103, 105, 115–117, 144
Fleischmann, M. 44, 93
Forman, M. 37
Fourier, J.B.J. 38
Fox Keller, E. 45, 76
Frankel, E. 36, 37
Franklin, R. 22, 81
Friedman, S.M. 123

Gage, M. 22
Gale, G. 45
Galilei, G. 126
Gall, F.J. 45
Garfinkel, H. 59, 62, 64
Gaskell, G. 111, 123
Gay-Lussac, J.L. 50
Gieryn, T. 106
Giglioli, P.P. 76
Gilbert, N. 29, 65
Goodel, R. 114
Gottlieb, M. 131, 132
Grasso, A. 123
Gregory, J. 21, 76, 118, 119, 145
Grmek, M.D. 123
Gross, P.R. 96
Grundmann, R. 118
Guillemin, R. 62
Guzzetti, L. 141

Hacking, I. 65, 105
Haeckel, E.H. 42
Hagstrom, W. 19, 20, 141
Haken, W. 101, 106
Hall, A.R. 23, 92
Halley, E. 78, 79
Hansen, A. 113, 114, 123
Haraway, D. 58, 139–140, 145
Harrison, J. 78, 79
Harvey, W. 44
Henderson, P. 113
Herrick, J. 117
Hess, D.J. 16, 21, 57
Hesse, M.J. 40
Hilgartner, S. 95, 123
Hobbes, T. 58, 96
Hood, L. 129
Hoyle, F. 33, 75, 76
Hughes, T.P. 81–82
Huxley, T.H. 42

Jacobi, D. 118
Jacques, J. 107
Jamison, A. 15

Kaufmann 101
Kevles, D.J. 43, 129
Kitzinger, J. 113, 123
Klaw, S. 7
Knorr Cetina, K. 63, 66, 144
Kohler, R.E., 38, 44
Kuhn, T.S. 26–29, 31, 32, 33, 38–39, 56, 73, 93, 94, 103, 122, 144

Lacan, J. 93
Lakatos, I. 53
Laplace, P.S. 36, 37, 41
Latour, B. 62, 63, 64, 70, 71, 72, 73, 74, 75, 78, 87, 93, 94, 96, 97, 102, 115, 122, 144
Layton, E.T. 79, 81
Le Verrier, U.J. 41
Levitt, N. 96
Lewenstein, B. 3, 19, 44, 108, 145
Lhuiler, S.A.J. 53
Liebig, J. von 50, 56
Lindemann, F.A. 14
Lupton, D. 112
Lyotard, J.F. 94

Index of names

Macdonald, S. 123
Mackenzie, D. 81, 82, 83, 88, 89, 92, 106
Mannheim, K. 2, 15–16
Marayama, S. 117
March, J.P. 106
Marx, K. 2
Matthiessen, L. 55
Maxwell, J.C. 12, 89
Mazzolini, R.G. 3, 44, 113, 123
Mazzotti, M. 3, 48
Mendel, E. 25, 43, 46
Merton, R.K. 3, 13–23, 42, 55, 58, 76, 77, 78, 93, 95, 99, 104, 105, 126, 135, 143
Miller, S. 19, 21, 76, 118, 119, 145
Millikan, R.A. 99, 103
Mitroff, I. 18, 19
Morell, J.B. 50
Mulkay, M. 19, 29, 65, 76

Nagel, T. 1
Nelkin, D. 123
Newton, I. 126
Nowotny, H. 3, 139, 143, 145

Olsen, J.P. 106
Ospovat, D. 43

Pascal, B. 21
Pasteur, L. 12, 32, 65, 70, 72–75, 97, 119
Pauling, L.C. 117
Pearson, K. 47
Penzias, A. 27, 81
Perec, G. 61, 93
Peters, H.P. 110, 123
Phillips, D.M. 113
Pickering, A. 36, 42, 144
Pinch, T. 66, 67, 68, 83, 84, 86, 88, 90, 104, 144
Pizzorno, A. 104
Planck, M. 32
Poisson, S.-D. 36, 37
Pons, S. 43, 93
Popper, K. 1, 29
Postol, T. 89, 90
Powell, W.W. 106
Price, D.J. de Solla 7–10, 21, 134, 143

Priestley, J. 52
Proust, M. 21
Pumfrey, S. 121

Raichvarg, D. 107
Raup, D. 76
Rayleigh, J.W.S. 20
River, C. 140
Roberts, L.W. 113
Roosevelt, F.D. 79
Rosen, P. 87–88
Rosenberg, N. 80, 83
Rositi, F. 123
Rossi, P. 23, 81
Rossiter, M. 22
Rousseau, J.J. 21
Roux, E. 73, 74
Rubin, V. 22
Rudwick, M.J. 45

Sarton, G. 16
Schaffer, S. 58
Schiaparelli, G.V. 49
Scrope, G.P. 45
Seagall, A. 113
Shapin, S. 41–44, 46, 48, 57, 58, 81, 144
Shinn, T. 114, 115, 118, 123
Silverstone, R. 123
Sinsheimer, R. 127
Snow, C.P. 14
Sobel, D. 79
Sokal, A. 93–96, 144
Sorokin, P.A. 16
Stendhal, H. 21

Taschwer, K. 3, 143
Temin, H.M. 43
Thomson, T. 50, 56
Thomson, J.J. 99
Tizard, H. 14

Van der Deder 61
Venter, C. 125, 129, 130, 131, 134
Virchow, R. 44, 50
Von Aitick 61

Wajcman, J. 81, 82, 83, 144
Watson, T.J. 22, 28, 82, 126, 128, 133

Weber, J. 20, 67, 69
Weber, M. 13
Webster, A. 134, 135
Wegener, P. 32, 52
Whevell, W. 12
Weinberg, A. 7
Weinberg, S. 94
White, L. 77, 92
Whitley, R. 117
Wilkie, T. 127–128, 133

Wilson, R. 27, 81
Wood, R. 51, 59
Woolgar, S. 58, 62–64, 144
Wynne, B. 37, 112, 123

Young, R.M. 45

Ziporyn, T. 109
Zloczower, A. 12
Zola, E. 21
Zuckerman, H. 3, 7, 19